Diversity in the Genus *Apis*

Studies in Insect Biology
Michael D. Breed, Series Editor

Diversity in the Genus Apis, edited by Deborah Roan Smith

Sweet Potato Pest Management: A Global Perspective, edited by Richard K. Jansson and Kandukuri V. Raman

Diversity in the Genus *Apis*

EDITED BY

Deborah Roan Smith

Routledge
Taylor & Francis Group

LONDON AND NEW YORK

First published 1991 by Westview Press, Inc.

Published 2018 by Routledge
52 Vanderbilt Avenue, New York, NY 10017
2 Park Square, Milton Park, Abingdon, Oxon OX14 4RN

Routledge is an imprint of the Taylor & Francis Group, an informa business

Library of Congress Cataloging-in-Publication Data
Smith, Deborah Roan.
  Diversity in the genus *Apis* / edited by Deborah Roan Smith.
    p. cm. — (Westview studies in insect biology)
  Includes bibliographical references and index.
  ISBN 0-8133-8057-X
  1. Apis (Insects) I. Title. II. Series
QL568.A6S65 1991
595.79′9—dc20   90-43610
                                                              CIP

ISBN 13: 978-0-367-01600-5 (hbk)
ISBN 13: 978-0-367-16587-1 (pbk)

To the memory of my mother

Ruth Evangeline Dix Smith,
11 November 1921 - 31 January 1989

Contents

Preface

Honey bees constitute a single genus in the family Apidae. One species, *Apis mellifera*, is native to Europe, Africa and the Middle East; the rest, a handful of species, are found in Asia. All honey bees are similar in morphology, social biology, nest architecture, foraging behavior, and the use by foragers of a complex "dance" to signal direction and distance to food sources. *Apis mellifera* is one of the best-studied insects in the world, though many basic questions about the biology of this species remain unanswered. However, the similarities among honey bee species have, to a certain extent, blinded us to the tremendous diversity of behavior and ecology found among the Asian species. Even the number of *Apis* species is not known with any certainty.

A major question in social biology, the single or multiple origins of highly eusocial behavior in the Apidae, is still hotly debated. Its resolution hinges on deciphering the phylogenetic relationships of the four apid subgroups (orchid bees, bumble bees, stingless bees and honey bees). This problem is addressed by three studies in this collection, using morphological data, nuclear DNA characters and mitochondrial DNA characters. The fact that no consensus is reached reflects the difficulty of the problem and shows the need for additional research.

This volume grew out of an informal conference, "Diversity in the Genus *Apis*," which took place in 1989 at the annual meeting of the Entomological Society of America. The conference was inspired largely by the work of Dr. Friedrich Ruttner, whose research on *Apis mellifera* spans many decades. In *The Biogeography and Taxonomy of Honey Bees* (1988) Ruttner reviewed and summarized published research on the comparative ecology, behavior, morphology and biogeography of *Apis mellifera* and the Asian honey bee species. The contrast between the voluminous literature on the western honey bee, *Apis mellifera,* and the relatively scanty data on the Asian honey bee species highlighted the need for further research on Asian *Apis*.

Conference participants were Fred C. Dyer (diversity in dance language), Thomas Seeley (comparative energetics in Asian *Apis*), Gudrun Koeniger (diversity in *Apis* mating systems), Jean-Marie Cornuet (genetic diversity in *Apis mellifera*), Gard W. Otis (isozyme variability in the genus *Apis*), Deborah Smith (mitochondrial DNA diversity in *Apis*) and Walter

S. Sheppard (ribosomal RNA diversity in *Apis*). Here, I have collected together (in expanded form) the papers presented by most of the original conference participants as well as additional chapters by other authors. These additional chapters include a discussion of the phylogenetic relationships among *Apis* species by Byron Alexander; a summary of species diversity in *Apis* by Gard Otis, which includes information on the biology of two newly recognized species, *A. andreniformis* and *A. koschevnikovi*; two studies of phylogenetic relationships within the Apidae by Sydney Cameron and Michael Prentice; and a synthesis of recent studies on honey bee energetics by Fred Dyer.

Together, these chapters present a well-rounded picture of current research on honey bee biology. These studies are for the most part very recent; in fact many are still in progress. They are meant to serve as introductions to various aspects of honey bee biology and as guides for future research, especially on Asian honey bees. In aid of this, each author has presented a discussion of the rationale for his or her study and the techniques and methodologies involved. I hope that this collection of chapters will be the starting point for many new research projects on the systematics, biogeography and comparative biology of the Asian honey bee.

Deborah Roan Smith

About the Contributors

Byron Alexander, Department of Entomology, Snow Entomological Museum, Snow Hall, The University of Kansas, Lawrence, KS 66045, United States

Sydney A. Cameron, Biology Department, Washington University, St. Louis, MO 63130, United States

Jean-Marie Cornuet, Laboratoire de Neurobiologie Comparée des Invertébrés, INRA-CNRS La Guyonnerie, 91440 Bures sur Yvette, France, and Laboratoire de Biologie et Génétique évolutives, CNRS, 91198 Gif sur Yvette, France

Fred C. Dyer, Department of Zoology, Michigan State University, East Lansing, MI 48824, United States

Gan Yik-Yuen, Jabatan Bioteknologi (Department of Biotechnology), Fakulti Sains Makanan dan Bioteknologi, Universiti Pertanian Malaysia, 43400 Serdang, Selangor, Malaysia

Lionel Garnery, Laboratoire de Neurobiologie Comparée des Invertébrés, INRA-CNRS La Guyonnerie, 91440 Bures sur Yvette, France, and Laboratoire de Biologie et Génétique évolutives, CNRS, 91198 Gif sur Yvette, France

Gudrun Koeniger, Institut für Bienenkunde, (Polytechnische Gesellschaft), Universität Frankfurt, FB Biologie, Karl-von-Frisch-Weg 2, D-6370 Oberursel, Germany

Bruce A. McPheron, Department of Entomology and Institute of Molecular Genetics, The Pennsylvania State University, University Park, PA 16802, United States

Makhdzir Mardan, Department of Plant Protection, Universiti Pertanian Malaysia, 43400 Serdang, Selangor, Malaysia

Gard W. Otis, Department of Environmental Biology, University of Guelph, Guelph, Ontario N1G 2W1, Canada

Michael Prentice, Department of Entomology, University of California, Berkeley, CA 94720, United States

Walter S. Sheppard, USDA-ARS, Bee Research Laboratory, BARC-East, Building 476, Beltsville, MD 20705, United States

Deborah Roan Smith, Museum of Zoology, Insect Division and Laboratory for Molecular Systematics, University of Michigan, Ann Arbor, MI 48109, United States (current address: Department of Entomology, University of Kansas, Lawrence, KS 66045, United States)

Tan S. G., Department of Biology, Universiti Pertanian Malaysia, 43400 Serdang, Selangor, Malaysia

1

A Cladistic Analysis
of the Genus *Apis*

Byron Alexander

Introduction:
Historical Review of Ideas About
Honey Bee Phylogeny

The literature on honey bees is so vast and incorporates contributions from such a wide range of different disciplines that few, if any, working scientists even pretend to be familiar with all of it (the author of this chapter certainly does not). Consequently, a given publication's pathways of influence are apt to be very strange and convoluted. As a case in point, consider a paper by A. Gerstäcker with the formidable nineteenth-century German title, "Über die geographische Verbreitung und die Abänderungen der Honigbiene nebst Bemerkungen über die ausländischen Honigbienen der alten Welt." This paper was apparently originally presented as a plenary address at a meeting of German beekeepers in Potsdam in 1862. It was printed in the form of a Festschrift, or commemorative volume, that also served as an identification badge indicating who had registered to attend the conference (Buttel-Reepen, 1906). Therefore, the number of copies originally printed was presumably only 524 -- the number of documented participants at the meeting -- and most of these copies were probably soon discarded (or perhaps converted to kindling for smokers to calm aggressive German honey bees). Nevertheless, this paper managed to form the groundwork for the prevailing orthodoxy concerning honey bee phylogeny for at least half a century. Gerstäcker presented a careful and comprehensive survey of variation in a number of morphological traits

1

across the entire geographic range of honey bees. He summarized his findings by recognizing only four species in the genus *Apis*. For the record, the names he used were *mellifica, dorsata, indica,* and *florea*; a detailed discussion of nomenclature is beyond the scope of this paper and can be found elsewhere (Maa, 1953; Ruttner, 1988)). Thus, the idea that the genus *Apis* consists of a small number of widespread species exhibiting considerable geographic variation considerably predates Mayr's (1942) formulation of the biological species concept. Gerstäcker also proposed splitting *Apis* into two informal groups, with his "First Group" comprised solely of *dorsata*, and his "Second Group" holding the other three species. He convincingly defended the merits of his system relative to a simpler one first proposed by Latreille in 1804 and adopted by Lepeletier in 1836, which grouped species solely on the basis of the color of the scutellum. An English version of Gerstäcker's work was published in 1863 in the *Annals and Magazine of Natural History*. Thus, Frederick Smith, a well-known English taxonomist at the British Museum (Natural History), was familiar with Gerstäcker's work when he published his own revision of honey bees in 1865. However, Gerstäcker's work was virtually unknown in Germany until it was resurrected in 1906 by von Buttel-Reepen, an influential German systematist and evolutionary biologist who was particularly interested in the evolution of social behavior in bees. An attentive official at the Berlin Zoological Museum had managed to acquire an original copy of Gerstäcker's 1862 Festschrift paper and showed it to von Buttel-Reepen, who was sufficiently impressed with it that he arranged to have it reprinted in its entirety as part of a comprehensive paper he published in 1906 on the systematics, biology, and biogeography of honey bees. As far as I have been able to determine (always keeping in mind the possibility of other undiscovered equivalents of Gerstäcker's papers), this 1906 publication by von Buttel-Reepen was the first serious attempt to discuss the evolution of the genus *Apis*, drawing on evidence from paleontology, biogeography, and comparative behavioral studies. It even included a phylogenetic diagram for all social Apidae (honey bees, stingless bees, and bumble bees) that not only indicated postulated ancestor-descendant relationships but also specified the geological age and current geographical distribution of each taxon. By our present standards, the justification for this elaborate phylogenetic scenario is extremely vague, so that it may seem like little more than grandiose story-telling. However, if considered in the context of the time when it was written, it represents an innovative and comprehensive attempt to provide a coherent conceptual framework for explaining observed patterns of variation among different species of honey bees, and it was certainly influential with other honey bee researchers. Furthermore, it was not entirely without logical or empirical support. To a large extent, von Buttel-Reepen based his proposed phylogeny of *Apis* on ideas about the most likely sequence of changes in the progressive development of social behavior. He considered the European honey bee to have the most advanced form of society, and *Apis dorsata* to be the least socially advanced. Specific features of *dorsata* behavior that he considered to be primitive included the rearing of workers and drones in cells of the same size, and a tendency to exhibit migratory behavior. The observation that meliponines also rear workers and drones in the same types of cells was cited as evidence that this is a primitive behavior in *dorsata*, since he considered the societies of meliponines to be less advanced than, and ancestral to, those of *Apis*.

Von Buttel-Reepen's ideas about honey bee phylogeny were generally either accepted or ignored, but not contested, for the next fifty years. When Maa published his comprehensive and extremely thorough taxonomic revision of honey bees in 1953, he had relatively little to say about phylogeny and did not challenge any of von Buttel-Reepen's major conclusions. It was Maa's opinion (1953, pp. 633-634) that "The genus *Megapis* [= *Apis dorsata*, or the *dorsata* species group] beyond doubt includes the most primitive forms. This assumption is fully supported both by morphological and biological facts. The relative positions of *Apis* [= *A. mellifera, A. cerana,* and *A. koschevnikovi*] and *Micrapis* [= *Apis florea* and *A. andreniformis*], however, are open to controversy." In his Table 6 (Maa, 1953, pp. 629-630), he presented a list of morphological characters "believed to be of phylogenetic significance," in which he contrasted "generalized" and "specialized" extremes of characters, although he presented no discussion at all of how he determined which extreme was generalized and which was specialized for each character.

An alternative proposal for the phylogeny of *Apis*, and the one most widely accepted today, grew out of Martin Lindauer's (1956) comparative study of the dance language in three species of *Apis* in Sri Lanka (formerly Ceylon). The explict goal of this study, which is now regarded as a classic work in comparative ethology, was to understand how the dance language of *Apis mellifera* could have evolved. Lindauer was the first to show that a dance language containing information about the direction and distance to resources occurs throughout the genus *Apis*. Although the basic form of the dance language was the same in all species he studied, he noted interspecific differences in details of the dance. He argued that this variation held clues to the evolution of the dance language, provided that one could identify homologous components of the language in each species and establish which condition was ancestral.

Lindauer's procedure for deciphering the history of the dance language was to analyze it from a functional perspective, under the assumption that a more direct, simpler communication system would be ancestral to a more indirect and complex system. He placed particular emphasis on the mode of communicating directional information, since the distinction between simpler and more complex signals seemed most obvious in this system. In *Apis florea*, the waggle dance is performed on a horizontal surface, and the straight run portion of the dance is pointed right in the direction foragers should fly when they leave the nest. The other honey bee species that Lindauer studied perform the dance on the vertical face of a honeycomb, and directional cues are presented with reference to the pull of gravity. Bees reading the dance must transpose these gravity-related directional cues in order to fly in the proper direction when they leave the nest. Lindauer reasoned that the method of indicating direction was clearly simpler and more direct, and thus more likely to represent the ancestral condition, when the dance was performed on a horizontal surface.

Lindauer's ideas gained wide acceptance, not only among German-speaking scientists, but also among English-speakers when presentations in English (Lindauer, 1961; von Frisch, 1967) became available. Additional evidence supporting Lindauer's evolutionary scenario was provided when the Janders conducted a comparative study of geotaxis in numerous families and genera of Malaysian bees. They found that the species of *Apis* that perform their dances on a vertical surface have a unique form of geotaxis, which they termed metageotaxis. This metageotaxis is essential for the accurate presentation of directional information in the waggle dance when it is performed on a vertical surface. Consequently, it seemed especially significant that the dwarf honey bee (the Janders were probably working with *Apis andreniformis*), which performs its dance on a horizontal surface, does not exhibit metageotaxis.

Biochemical data presented by Kreil (1973, 1975) have also been interpreted as supporting Lindauer's hypothesized phylogeny (Kreil, 1975; Ruttner, 1988). Kreil determined the complete sequence of 26 amino acids in the peptide melittin from the venom glands of four species of *Apis*. He found that the sequence was identical in *mellifera* and *cerana*, and *florea* was least like the other three species.

The first author to call attention to uncertainty about the polarity (i.e., primitive vs. derived condition) of various characters in Lindauer's scenario was N. Koeniger (1976). He pointed out that data such as Kreil's provided information about phenetic similarity, but were not informative about phylogenetic relationships in the absence of information about character polarity (Appendix 1 contains a detailed explanation of the ambiguity of Kreil's data). Koeniger also called attention to further studies of geotaxis by Horn (1975) which raised doubts about whether the form of geotaxis in *Apis florea* was the same as that of other bees. The Janders had shown that *Apis florea*, along with all the other bees they examined except *Apis cerana*, *dorsata*, and *mellifera*, has a form of geotaxis that they called progeotaxis. Horn's work indicated that, although the form of the behavior called progeotaxis was the same in *Bombus terrestris* and *Apis florea*, the sensory receptors mediating the behavior are different. Without further study in the outgroup, it is not clear whether the geotaxis of *Apis florea* is plesiomorphic or an autapomorphy, perhaps derived from the metageotaxis of other *Apis* species.

Furthermore, Koeniger pointed out that *Apis* belongs to a monophyletic group (the family Apidae of Michener, 1974) in which cavity-nesting is probably the groundplan state. Hence, the cavity-nesting of certain honey bee species may not be a derived condition, as Lindauer's evolutionary hypothesis requires. Taken by itself, this character suggests that the species that nest in more open situations may have arisen later than the cavity-nesting honey bees. Dyer (1985, 1987) has recently published detailed analyses of the dance language in *A. florea* that show that its method of communicating directional information is not as simple or straightforward as Lindauer's more cursory studies had suggested.

Although Koeniger called attention to the need for a rigorous examination of characters in the light of the principles of phylogenetic systematics (Hennig 1950, 1966), and he discussed how a cladistic interpretation of one character would alter widely-held opinions about *Apis* phylogeny, he did not present a comprehensive cladistic analysis himself. Instead, he ended his paper with the statement (my translation) that "a clarification [of the phylogeny] will require studies involving characters of *Apis* species that can be compared with homologous characters of other groups of Apidae." To my knowledge, the only previously published study of *Apis* claiming to be a quantitative cladistic analysis is that of Sakai et al. (1986). They evaluate 23 species and subspecies (or races) from the entire natural range of the genus. Their data matrix includes morphological, behavioral, and biochemical characters, and they present both cladistic and phenetic analyses. However, their cladistic analysis is difficult to understand or evaluate, because they do not indicate how the characters were polarized or how coding decisions were made for continuously varying characters. Nevertheless, their conclusions about cladistic relationships are concordant with Lindauer's hypothesis.

In view of Koeniger's (1976) remarks about the need for additional characters whose polarities could be clearly established, it is ironic that he must have been unaware that just such a set of characters had been discovered by Snodgrass in 1941, in a comparative morphological study of the male genitalia throughout the order Hymenoptera. In this wide-ranging survey, Snodgrass examined three species of *Apis* (*mellifera*, *cerana* (which he called *indica*), and *florea*), identified several similarities shared by *mellifera* and *cerana*, and presented cogent arguments indicating that they were *derived* homologous similarities.

The analysis presented in this chapter is an extension of the study begun by Snodgrass and Koeniger. It is a quantitative cladistic analysis based upon a comparative study of adult morphology and, to a much smaller extent, behavior. Its objective is to determine if a cladistic analysis, using characters whose polarity can be established by outgroup comparison and which are independent of those involved in Lindauer's hypothesis of the evolution of the dance language, will support the cladistic relationships among the species of *Apis* implied by his evolutionary scenario. My data matrix (Table 1.2) does include one character that is clearly an integral part of Lindauer's evolutionary scenario. This character is the choice of nest site location (character number 20 in Tables 1.1 and 1.2). I included it in my analysis because Koeniger has made a convincing case that its polarity can be determined by outgroup comparison, and because he specifically mentioned it as supporting an alternative phylogenetic hypothesis for the species of *Apis*. In reporting my results, I have tried to explain the meaning of specialized terminology and quantitative measurements that are unlikely to be familiar to readers not versed in the methodology of quantitative cladistic analysis. However, a complete explanation and justification of parsimony methods is beyond the scope of this chapter. An introduction to the philosophical, biological, and mathematical issues involved in parsimony methods as used in phylogenetic systematics can be found in Farris (1983), Felsenstein (1983), and Sober (1983).

Materials and Methods

Before a cladistic analysis of the species of *Apis* can be undertaken, it is necessary to establish what the species are. Taxonomists working on honey bees have differed widely in the number of species they recognize. Recent treatments have ranged from Maa's (1953) upper extreme of 24 species in three genera to a much more conservative list of four, or possibly five, species in one genus (Ruttner, 1988). However, until very recently, the prevailing practice was to ignore Maa's work entirely or dismiss it as an example of extreme splitting, and recognize four species of *Apis*, namely *florea, dorsata, cerana*, and *mellifera*, following the practice first championed by Gerstäcker. Even the most recent general summaries on honey bee biology or taxonomy (e.g. Seeley 1985, Winston 1987, Ruttner 1988) follow this practice, with some brief discussion of a few unresolved taxonomic questions concerning certain Asian populations. Serious consideration of these unresolved questions is now under way, with the result that opinions about the number of species of *Apis*, especially in Southeast Asia, are changing. The analysis presented here deals with six species or species groups: *andreniformis, cerana, florea, koschevnikovi, mellifera*, and the "*dorsata* group". Evidence for the recognition of *andreniformis* and *koschevnikovi* as valid biological species is discussed elsewhere (Wu and Kuang, 1986, 1987; Wongsiri et al., 1989 for *andreniformis*; Tingek et al., 1988; Mathew and Mathew, 1988; Rinderer et al., 1989; Ruttner et al., 1989 for *koschevnikovi*). The species status of various populations of *Apis* with obvious affinities to *dorsata* remains more controversial. The cladistic analysis presented here is not intended to resolve whether *dorsata* is one widespread species exhibiting considerable geographic variation (similar to that of *cerana* and *mellifera*), or a group of closely related species. Cladistic relationships among taxa can be resolved only if one can identify characters that exhibit more than one state in the taxa being compared. The characters that could be used to resolve relationships among the six taxa analyzed in this study did not vary within the *dorsata* group. If systematists studying *Apis* reach a consensus that there are several species in the *dorsata* group, it will be necessary to find additional characters to resolve the cladistic relationships among them. The resolution of this study will be sufficient to provide an independent test of the hypothesized phylogeny of the species studied by Lindauer, and it can provide a framework for future studies using other characters that might provide a finer level of resolution.

In selecting characters for this analysis, the principal criteria were that there be two or more discrete states for each character, and that one of these states also occur in the outgroup (i.e. species outside the genus *Apis*, but considered to be close relatives of *Apis*), so that there would be a basis for determining which of the alternate states in *Apis* represents the ancestral condition. Care was taken to examine specimens from throughout the known range of each species, to verify that the characters chosen are not restricted to local populations of a species. Although all castes have been examined, drones were found to have most of the variation that was potentially informative about phylogenetic relationships within the genus. Outgroup taxa used for character polarization were chosen from all the major lineages in the family Apidae (*sensu* Michener, 1974), whose monophyly is well established (Winston and Michener, 1977; Sakagami and Michener, 1987), and from two basal clades within the Xylocopinae, which is the sister group of the Apidae (Sakagami and Michener, 1987). Additional details of the taxa examined in this study are documented elsewhere (Alexander, 1991). Although several cladistic analyses of the family Apidae have been published (Winston and Michener, 1977; Kimsey, 1984; Plant and Paulus, 1987; Chapters 3, 4 and 5, this volume), there is still uncertainty as to the sister group of *Apis* (C.D. Michener, personal communication). This did not present a problem with most of the characters used in this study, since the outgroup exhibited only one of the states found within *Apis*, so that polarity decisions were unequivocal (*sensu* Maddison et al., 1984). The hind wing venational character for which this was a significant consideration will be discussed below.

Dissections of male genitalia and female stings were cleared overnight at room temperature in 10% KOH. These dissections are in vials mounted with the pinned, dried specimens from which they were dissected. Specimens with dissections that were examined for this study are identified as voucher specimens in the collections of the Snow Entomological Museum and the Cornell University Insect Collections. Cladistic analyses were done with the Hennig86 computer program written by James S. Farris (Farris, 1988).

Results

The analysis used 21 characters (Table 1.1; matrix in Table 1.2) and 7 taxa, including the outgroup. The general objective of a cladistic analysis based upon parsimony methods is to arrange taxa in a pattern that minimizes the number of transitions among character states. In biological terms, parsimony can be viewed as an attempt to maximize the number of shared similarities that can be explained by homology, or inheritance from a common ancestor, while minimizing the number of similarities that must be explained by *ad hoc* hypotheses of homoplasy (= convergence and parallelism) (Farris, 1983). In mathematical terms, this amounts to finding the shortest possible pathway, or tree, connecting the taxa. Tree "length" is expressed as the number of "steps", with a step being a transition from one character state to another. Consequently, it is important to specify how characters that exhibit more than two states, such as characters 11, 12, 13, and 17 in Table 1.1, are to be treated in the algorithms used to find the shortest tree. If a character is treated as *additive*, a transition between states 0 and 2 must proceed through state 1, so that the transition between states 0 and 2 requires two steps. If a character is treated as *non-additive*, the transition between states 0 and 2 need not pass through state 1, and any transition between any two character states requires only one step. Non-additive characters place fewer restrictions on character transformations, but they also have less power to resolve cladistic relationships (Mickevich, 1982). Thus, in an analysis aimed at resolving cladistic relationships, it is generally preferable to treat multistate characters as additive whenever possible. In this analysis, only character 17, the vestiture of the male tarsi, was coded as non-additive, since there was no independent biological rationale (such as a morphocline) for arranging the three alternate states in a linear transformation series. The other multistate characters (11-13) were coded as additive, since one could logically consider them to be ordered in a linear sequence of transformations (or morphocline), with the starting point, or plesiomorphic condition, determined by noting which character state occurs in the outgroup.

A parsimony analysis of the matrix in Table 1.2 found a single most parsimonious tree (Figure 1.1), with a length of 27 steps and a consistency index of 93. The consistency index for a tree is the ratio of the shortest possible tree for a data set (its length if *all* shared similarities are homologous and *no* character states arise independently in unrelated taxa) to the observed length of the tree whose consistency index is being calculated, taking into account multiple origins of a given character state on that tree (Kluge and Farris, 1969). The higher the consistency index, the greater the agreement among different characters in supporting the same pattern of cladistic relationships. Compared to other published quantitative cladistic analyses, the consistency index of 93 found in this analysis is unusually high.

One possible reason for such a high consistency index is that over half of the characters in the data matrix are synapomorphies shared by *all* species of *Apis*. Such characters support the non-controversial hypothesis that honey bees are a monophyletic assemblage. (Although this hypothesis is not controversial, it is not valid to simply assume that *Apis* is monophyletic.) However, these characters supporting the monophyly of *Apis* provide no information at all about the question of major interest in this study, namely the phylogenetic relationships among the species *within* the genus. If one is to use the consistency index to assess how much agreement or disagreement there is among different characters in supporting the same pattern of phylogenetic relationships, characters that

Table 1.1 Characters and alternate states used in the quantitative cladistic analysis. Characters 19 and 20 refer to the behavior of workers or larvae; all other characters are features of adult morphology. Characters 0-3 apply to both sexes and all castes, 4 applies to queens, 5-7 apply to workers, and 8-18 apply to drones. All characters except 17 are coded as additive (see text). Alexander (1991) contains illustrations of Characters 0 and 5-9, which are not illustrated here or in other references cited below.

ʹ

CHARACTERS AND CHARACTER STATES

0. Compound eyes hairy:
   0. no
   1. yes
1. Angle ABC of forewing:
   0. > 45° (Figure 1.2D)
   1. < 45° (Figure 1.2A)
2. Angle BDE of forewing:
   0. > 45° (Figure 1.2D)
   1. < 45° (Figure 1.2A)
3. Distal abscissa of hindwing vein M (indica vein):
   0. present (Figure 1.2C)
   1. absent (Figure 1.2B)
4. Ovariole number:
   0. 3 or 4*
   1. > 50
5. Barbed sting:
   0. absent
   1. present
6. Sting sheath:
   0. pigmented and bearing distinct setae
   1. unpigmented, with short, inconspicuous setae
7. Venter of metasomal segment 8 a conspicuous membranous bulb surrounding base of sting shaft:
   0. no
   1. yes
8. Compound eyes of males meeting at top of head
   0. no
   1. yes
9. Male proboscis:
   0. same length as in female (worker in social species)
   1. much shorter than in worker
10. Male endophallus (Figure 3.5 in Ruttner 1988):
    0. not greatly enlarged
    1. enormously enlarged

(continues)

Table 1.1, continued

CHARACTERS AND CHARACTER STATES

11. Ventral gonocoxite:
    0. present, sclerotized throughout
    1. sclerotized portion reduced to a transverse bar (Figure 1.3A, gc)
    2. membranous throughout (Figure 1.3C)
12. Dorsal gonocoxite:
    0. not conspicuously reduced (Figure 1.4C, gc)
    1. reduced, widely separated mesally, about half as long as penis valves (Figure 1.4E, gc)
    2. greatly reduced, less than half as long as penis valves (Figure 1.4D, gc)
13. Gonobase:
    0. present as a distinct ring
    1. an incomplete ring, or isolated fragments of sclerotization (Figure 1.4C, br)
    2. absent (Figure 1.4D, E)
14. Male metasomal tergum 8:
    0. with two long arms of about the same length (Figure 1.3A, C, T8)
    1. vertical arm much longer than horizontal arm (Figure 1.4A, T8; also Plate 31T of Snodgrass 1941)
15. Male metasomal sterna 7 & 8
    0. not fused mesally (Figure 1.3A-D, S7, S8)
    1. fused mesally (Figures 1.4A, B, S7, S8)
16. Thumblike process on male hind basitarsus:
    0. absent
    1. present (Figure 7.4 in Ruttner, 1988; Figures 1 & 2 in Wu and Kuang, 1987)
17. Vestiture of male tarsi:
    0. not specially modified
    1. dense pads of frond-like setae on middle and hind tarsi (Figures 8.8, 8.9 in Ruttner, 1988)
    2. dense pads of stiff bristles on inner surface of thumblike process of hind basitarsus (Figure 7.5 in Ruttner, 1988)
18. Flagellum of male antenna:
    0. "long", i.e. about as long as distance from vertex to apical margin of clypeus
    1. "short", i.e. about half as long as distance from vertex to apex of clypeus
19. Capping of drone cells:
    0. without a central pore
    1. with a central pore (Figures 9.13, 9.14 in Ruttner, 1988)
20. Nest Site
    0. within a cavity
    1. not within a cavity

*In the outgroup taxa examined for this analysis, the socially parasitic genus *Psithyrus* shows considerable intra- and interspecific variability in ovariole number (Cumber, 1949), although the number of ovarioles never approaches that found in *Apis* queens. This is presumably an autapomorphy for *Psithyrus*, associated with its socially parasitic way of life.

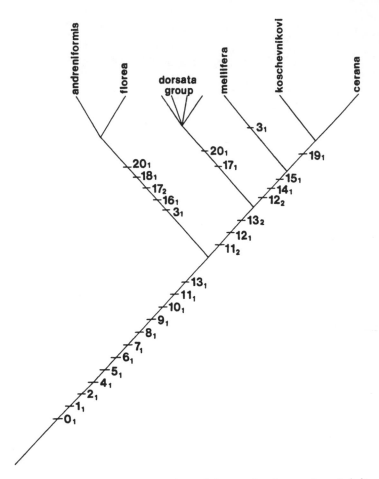

Figure 1.1 (A) The most parsimonious cladogram for the species of *Apis*. This cladogram is based upon the data matrix in Table 1.2, and explanations of characters and character states are given in Table 1.1. The pairs of numbers on the cladogram indicate inferred transformations in character states. In each pair, the large number on the left is the number of the character, as listed in Table 1.1, and the smaller subscript on the right represents the derived character state. See text for further discussion.

have the same derived state in all taxa in the group of interest, or characters in which the derived state occurs in only one taxon, will inflate the consistency index in a potentially misleading way. No matter how the taxa are grouped on a cladogram, these particular characters will never suggest an arrangement of taxa that would contradict the grouping of taxa supported by another character.

If the consistency index is recalculated with all the autapomorphies for the genus *Apis* excluded (characters 0-2 and 5-10), its value drops from 93 to 88. The only homoplastic (convergent) characters are the distal abscissa

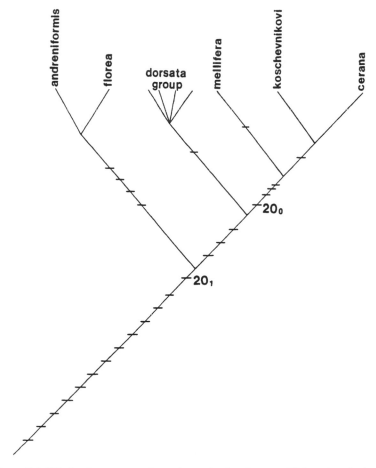

Figure 1.1 (B) An alternate transformation series for character 20 (nest location). See text for further discussion.

of vein M on the hind wing (character 3), which is hypothesized to have been independently lost in *mellifera* and the *andreniformis-florea* lineage, and nest site location, which will be discussed below.

Discussion

The results of this analysis are unusually clear and unequivocal. This does not guarantee that the hypothesis of common ancestor relationships summarized in Figure 1.1 is true, but the available evidence strongly favors this hypothesis over any others. The hypotheses of homoplasy required by this phylogenetic hypothesis are plausible. Two equally parsimonious transformation series, both involving homoplasy (Figure 1.1), could account for the distribution of Character #20 (nest site location). This character will be considered in more detail below, in the context of a discussion of the evolution of the dance language. The other character requiring an ad-hoc hypothesis of homoplasy is a feature of wing venation, the "indica vein," that has long been used to distinguish between *cerana* and *mellifera*, since it is consistently present in *cerana* (Figure 1.2 C) and absent in *mellifera* (Figure 1.2 B). Ruttner (1988, Table 1.1) reported this vein to be present in *dorsata* (including *laboriosa*), and "variable" in *florea*. He did not indicate how many specimens were examined, or whether the entire geographic range of each species was covered, in preparing Table 1.1 in his book (although his reputation for exemplary thoroughness and precision in such matters is certainly well deserved). I found this character to be variable in all *Apis* species, occasionally even between the two wings of a single individual bee. Furthermore, the distinction between its presence and absence is not absolute. Even when present, it is not a fully developed, tubular vein, but only a pigmented line (= nebulous vein, in the terminology of Mason, 1986), and its length and degree of pigmentation vary considerably. However, the frequency of aberrant individuals, relevant to the norm for a given species, was very low in all species, and not obviously higher in *florea* or *andreniformis* than in other species. As Ruttner has been correctly emphasizing for many years, proper evaluation of quantitative character variation in *Apis* requires samples of numerous colonies across the entire geographic range of a species. I did not have access to such samples for all species in this study. At present, it is only possible to make the qualitative statement that this character exhibits low levels of variation in all species of *Apis*. This wing venation character also varies among genera and species, and to an unmeasured extent within species, in the outgroup. In view of these observations, and of the nature of the other similarities shared by the *dorsata* group, *cerana*, *koschevnikovi*, and *mellifera* (summarized on Figure 1.1), I conclude that the latter similarities are truly homologous, and that two lineages of *Apis* have independently lost the terminal trace of a vein on the hind wing.

Additional characters amenable to cladistic analysis should be sought in order to determine if they also support the phylogenetic hypothesis presented here. In particular, Kreil's (1973, 1975) data on variation in the melittin molecule can very readily be analyzed if character polarities can be established by outgroup comparison. Other studies based on molecular data are currently underway, and the results of these studies are awaited with great interest. Although the monophyly of *Apis* is very strongly

15

Table 1.2. Data matrix for the quantitative cladistic analysis. Characters and character states are defined in Table 1.1. *"dorsata"* represents the entire *dorsata* group, which includes *dorsata, laboriosa, binghami,* and *breviligula.* A "?" in a cell in the matrix indicates that the character state is unknown for the taxon.

CHARACTERS

TAXA:	0	1	2	3	4	5	6	7	8	9	10	11	12	13	14	15	16	17	18	19	20
Outgroup	0	0	0	0	0	0	0	0	0	0	0	0	0	0	0	0	0	0	0	0	0
florea	1	1	1	1	1	1	1	1	1	1	1	1	0	1	0	0	1	2	1	0	1
andreniformis	1	1	1	0	1	1	1	1	1	1	1	1	0	1	0	0	1	2	1	0	1
dorsata	1	1	1	1	1	1	1	1	1	1	1	2	1	2	0	0	0	1	0	0	1
cerana	1	1	1	0	1	1	1	1	1	1	1	2	2	2	1	1	0	0	0	1	0
koschevnikovi	1	1	1	0	?	1	1	1	1	1	1	2	2	2	1	1	0	0	0	1	0
mellifera	1	1	1	1	1	1	1	1	1	1	1	2	2	2	1	1	0	0	0	0	0

supported by numerous independent morphological characters, some of the major groupings within the genus depend heavily upon interpretations of features of the male genitalia.

The latter point is particularly important because some of my character interpretations differ from those of other authors. In a comparative study of male and female genitalia in the Hymenoptera, Smith (1970) analyzed *Apis mellifera* and reached different conclusions from Snodgrass (1941, 1956) concerning the homology of the gonostylus and penis valves. Smith apparently did not examine any *Apis* species other than *mellifera*, nor did he explain why his interpretations differed from Snodgrass'. I accept the latter author's interpretations (presented on pages 62-63 and plates 31-33 of Snodgrass, 1941). Not only is the relative position of the structures that Snodgrass considers to be the penis valves exactly what one would expect from a comparison with other Hymenoptera, but these structures possess apodemes for muscle attachments (Figures 1.4 C,D,E, ap) that also occur on penis valves in other Hymenoptera. Smith considers these structures to be gonostyli, but gonostyli on other bees do not have such apodemes, and articulate directly with the gonocoxites (or are fused with the gonocoxites).

The cladistic implications of Smith's alternative interpretation are negligible for one character, but significant for the other. Whatever the true homology of the structures that Smith called gonostyli, they have the same basic form throughout the genus, and thus have no bearing on the cladistic analysis. However, Smith's interpretation of this character is tied to his interpretation of some sclerotized structures within the penis bulb of the enlarged endophallus of *Apis mellifera*. It is not clear from Smith's discussion if he was aware that these structures are found only in *mellifera*, and not in other species of *Apis*. He considers them to be the penis valves. (He actually calls them "gonapophyses IX", but it is clear (e.g. in Figure 11A of his paper) that he intends this term to be synonymous with Snodgrass' term "penis valves"). If Smith is correct, the penis valves are not only highly modified and physically displaced in a very unusual manner, but they suggest either a different interpretation of phylogenetic relationships among the species of *Apis*, or a peculiar evolutionary history for the penis valves in the genus. Taken by themselves, the penis valves as interpreted by Smith would suggest that *mellifera* is the basal lineage in *Apis*. Since it has the valves and no other species do, the simplest explanation would be that *mellifera* has retained the ancestral structure (albeit in a highly modified form), which was then lost in the common ancestor of the rest of the species of *Apis*. However, if one accepts this, one must also conclude that the shared similarities in the gonocoxites,

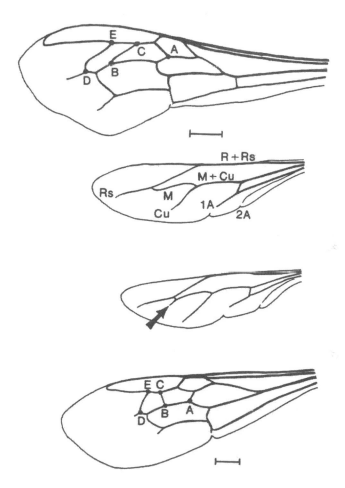

Figure 1.2 (A) Front wing of *Apis mellifera* worker. Letters refer to landmark points used in defining the angles referred to under Characters 1 and 2 of Table 1.1. Scale line = 1.0 mm; **(B)** Hind wing of *Apis mellifera* worker. Abbreviations for vein names (Comstock-Needham system) are as follows: R = radius, Rs = Radial sector, M = media, Cu = cubitus, 1A = first anal, 2A = second anal. Scale same as in A; **(C)** Hind wing of *Apis cerana* worker. Arrow indicates distal abscissa of vein M, or the "indica vein". Scale as in A; **(D)** Front wing of *Bombus pratorum*. Letters refer to landmarks as explained above in A. Scale line = 1.0 mm.

gonostyli and metasomal sterna 7 and 8 (Characters 11-15 in Tables 1.1 and 1.2) in *mellifera* and various other species are either plesiomorphic or were independently acquired in *mellifera* and other species. Comparison with the outgroup casts doubt on the interpretation that the similarities are plesiomorphic, so one must resort to numerous *ad hoc* hypotheses of homoplasy. Alternatively, one could conclude that the placement of *mellifera* on the cladogram of Figure 1.1 is correct, but that the penis valves were lost in the ancestral *Apis*, and somehow reappeared in *mellifera*, in a modified form and in a unique position *within* the endophallus. Or one could conclude that Smith misinterpreted the homologies of the gonostyli and penis valves in *Apis* as a consequence of examining only one species in the genus, and the sclerotized structures in the penis bulb of the endophallus of *mellifera* are not penis valves at all. I favor the last interpretation.

I differ from Snodgrass (1941) in my interpretation of one character of male genitalia, although in this case either interpretation will the support the same conclusion about cladistic relationships. Our difference of opinion concerns male metasomal sternum 8 (S8, Character 15 in Table 1.1). (Snodgrass used a numbering system that counted the first abdominal segment, which is fused with the thorax in apocritan Hymenoptera, as segment 1. Thus, segments referred to as metasomal segments 7 and 8 in most taxonomic works on bees are labelled as abdominal segments VIII and IX in Snodgrass' paper. To facilitate comparison with Snodgrass' figures, his numbers will be shown in parentheses in the following discussion, with S for sternum and T for tergum).

Snodgrass examined only three species of *Apis*: *florea*, *cerana*, and *mellifera*. In *florea*, S8 (IXS) is very narrow, and it articulates with T8 (IXT) laterally (Plate 31L in Snodgrass, 1941). In *cerana* and *mellifera*, the structure that Snodgrass interpreted as S8 is much broader ventrally, but it fuses laterally with the remnants of the dorsal gonocoxite, and does not articulate with T8 (IXT, Plate 31 N in Snodgrass, 1941).

A comparison of all the species of *Apis* suggests an alternative interpretation of the states in *mellifera*, *cerana*, and *koschevnikovi*. I do not consider S8 in these species to be a broad ventral band of sclerotization that fuses laterally with the gonocoxites. Instead, I consider that S7 and S8 are fused into a single sclerite that is greatly attenuated mesally. The broad sclerotized band that Snodgrass interpreted as S8 is either a part of the penis valves or a new sclerite not homologous with the metasomal sterna of other bees. Although some of the specimens that I have examined have a broad ventral band distinct from the penis valves, and thus resemble the condition illustrated by Snodgrass, I have also seen

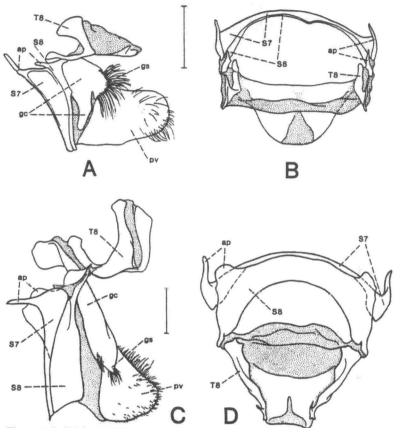

Figure 1.3 (A) Lateral view of genital capsule (excluding endophallus) and associated sterna and terga of male *Apis andreniformis* from the Malay Peninsula; ap = apodeme; gc = gonocoxite (= lamina parameralis of Snodgrass, 1941); gs = gonostylus (= paramere of Snodgrass, 1941); pv = penis valve (= gonapophyses IX of Smith, 1970); S7 = metasomal sternum 7 (= VIIIS of Snodgrass, 1941); S8 = metasomal sternum 8 (= IXS of Snodgrass, 1941); T8 = metasomal tergum 8 (= IXT of Snodgrass, 1941). Scale line = 0.5 mm. (B) Oblique dorsal view of metasomal sternum 7 and segment 8 in male *Apis andreniformis* from the Malay Peninsula; abbreviations and scale line as in Figure 1.4 B. The partially sclerotized dorsal region between the lateral sclerites labelled as T8 probably represents metasomal segment 9 (labelled as subanal plate of abdominal segment X in Snodgrass, 1941). This figure (as well as Figures 1.3 D and 1.4 A-E), is oriented with anterior to the top of the page, which is the same orientation used by Snodgrass (1941), but opposite to the conventional orientation in most illustrations of bee genitalia in taxonomic publications. (C) Lateral view of genital capsule (excluding endophallus) and associated sterna and terga of male *Apis dorsata* from the Malay Peninsula. Abbreviations as in A; scale line = 1.0 mm. (D) Oblique dorsal view of metasomal sternum 7 and segment 8 of male *Apis dorsata* from the Malay Peninsula; abbreviations and explanatory comments as in B; scale line as in C.

specimens in which this band is partially or completely fused with the penis valves (my Figure 1.4 A illustrates such a specimen). The evidence for a fusion of S7 and S8 in the cavity-nesting *Apis* is as follows.

In all *Apis* species, S7 is a very narrow band mesally and somewhat broader laterally, with very narrow, anteriorly-projecting apodemes arising from the anterolateral corners. In *florea* and *andreniformis*, S8 is very similar to S7, except that the anterolateral apodemes are much shorter (Figure 1.3 A,B, ap). S7 and S8 lie parallel to one another, with only minimal overlap laterally.

In the *dorsata* group, S7 is very similar to the corresponding sclerite in *florea*, but S8 is considerably broader, and the anterolateral apodeme is very broad and flat, rather than a narrow, fingerlike process (Figures 1.3 C,D, ap). The two sterna broadly overlap laterally, with S8 lying mesad of S7. A posterolateral process on S8 articulates with T8. Mesally the two sterna are distinct, parallel bands, with S8 much broader than S7.

As Snodgrass noted, S7 and S8 (VIIIS and IXS) are identical in *mellifera* and *cerana* (the same is true of *koschevnikovi*). The sterna are illustrated in Figure 1.4 A and B of this chapter, and in Snodgrass' Plate 31, Figures N and O, and all of Plate 33, but especially Figures C and D. If Snodgrass has interpreted S8 correctly for this group, his S7 (VIIIS) has some unusual structures laterally that are not present in other species of *Apis*. The characteristic narrow, fingerlike anterolateral apodemes are present, as in other species, but posterad of them is a distinct indentation and an anterior and posterior apodeme (shown most clearly in Figure 1.4 A and B of this chapter and Snodgrass' Plate 33, Figures C and D). It is particularly noteworthy that the posterior apodeme articulates with T8, as does the posterolateral apodeme of S8 in other species of *Apis*. Although the anterolateral apodeme is not identical in shape to corresponding apodemes on S8 of the *florea* or *dorsata* clades, it is sufficiently similar that it is not difficult to picture it as homologous. If these processes are in fact homologous to the corresponding structures on S8 of other *Apis* species, the inference is that S7 and S8 are fused in *mellifera*, *cerana*, and *koschevnikovi*. Laterally the fusion is not complete, so that the apodemes of the two sterna are still discernible. Mesally the combined sterna are greatly attenuated, and/or one is lost.

Under this interpretation, a *mellifera*-type fusion of S7 and S8 could have been derived from either a *florea*-like or a *dorsata*-like ancestor. The particular feature of the *mellifera*-type character state that is clearly apomorphic relative to either of the other types is the fusion of S7 and S8, and this is the distinction that was used in the quantitative cladistic analysis (Character 15 of Tables 1.1 and 1.2). In terms of the cladistic

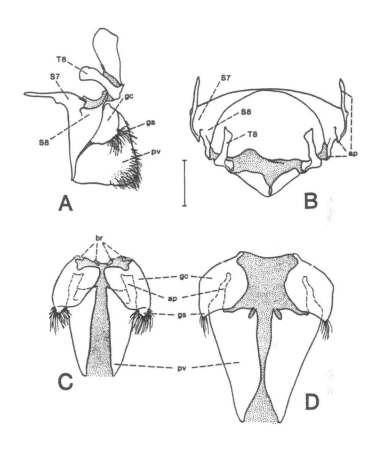

Figure 1.4 (A) Lateral view of genital capsule (excluding endophallus) and associated sterna and terga of male *Apis mellifera* from North America (New York); abbreviations as in Figure 1.3 A; scale line = 1.0 mm. (B) Oblique dorsal view of metasoma sternum 7 and segment 8 in male *Apis mellifera* from New York; abbreviations and explanatory comments as in Figure 1.3 B; scale line as in A. (C) Dorsal view of genital capsule (excluding endophallus) of male *Apis andreniformis* from the Malay Peninsula; br = sclerotized remnants of basal ring (= gonobase); other abbreviations as in Figure 1.3 A; scale line for C-E = 0.5 mm. (D) Dorsal view of genital capsule (excluding endophallus) of male *Apis mellifera* from New York; Abbreviations as in Figure 1.3 A;

analysis, either Snodgrass' interpretation of these structures or mine would lead to the same cladistic inference: *mellifera, cerana,* and *koschevnikovi* share a synapomorphy not found in other *Apis* species.

If Figure 1.1 represents a well-supported hypothesis of phylogenetic relationships among the extant species *Apis*, how does this influence our interpretation of the evolution of some of the unique features of honey bees? Since the phylogenetic hypothesis presented here does not differ from Lindauer's, except for the addition of some species he did not recognize, it may seem that it does not compel us to radically alter our thinking about the evolution of the dance language. However, it is important to note that the basal position of the clade *florea* + *andreniformis* on the cladogram does not necessarily mean that the dance language or nest architecture in these species is most like that of the ancestral *Apis*. Basal position on a cladogram does not necessarily mean that a lineage is "primitive". For example, the drones of *florea* and *andreniformis* have a greater number of unquestionably derived external morphological features (antennae, hind basitarsi) than any other species in the genus. In terms of external morphology (not genitalia) and overall appearance, the drones of the ancestral *Apis* may have looked more like *mellifera* than *florea*. Determining the polarity of these external morphological characters of drones is relatively straightforward, because outgroup comparison can clearly establish the plesiomorphic condition. Characters involving the dance language and nest architecture are more difficult to polarize, because it is often difficult or impossible to confidently identify in the outgroup *any* of the states found in *Apis*.

Lindauer took the dance language of *florea* to be a prototype of the ancestral *Apis*. More recent workers (Koeniger, 1976; Dyer, 1985, 1987) have questioned the validity of this assumption. What light does a cladistic analysis shed on this problem? One can use a cladogram as a basis for postulating the most parsimonious sequence of evolutionary transformations for any particular character of interest.

Consider, for example, an important element in Lindauer's scenario, the proposal that the ancestral *Apis* danced on a horizontal surface. Extant species exhibit two different states that are clearly tied to two basic types of nest architecture. In the majority of species, the constructed part of the nest consists entirely of vertical sheets of horizontally-directed, hexagonal waxen cells, and the bees dance on the vertical face of the honeycomb. In *florea* and *andreniformis*, the basic nest construct is the same, but the upper rows of cells are greatly lengthened and form the foundation for an expanded horizontal platform at the top of the nest. Dances are performed on this horizontal platform (although Dyer 1985, 1987 reports that they often begin or end below the top of this platform). Unfortunately, neither of these nest architectures occurs in the outgroup. A few meliponines have nests with vertical combs, but this is almost certainly an independently derived condition (Michener, 1974, 1990). Consequently, it is not clear which of the two types of dancing platforms within *Apis* represents the ancestral condition. If dancing on a horizontal surface is plesiomorphic, then one need only postulate a single transition to a vertical dancing surface, in the common ancestor of the *dorsata* group and the cavity-nesting species. If vertical dancing is the plesiomorphic condition, then one transition to horizontal dancing took place in the common ancestor of *florea* and *andreniformis*.

As noted earlier, Koeniger (1976) placed special emphasis on nest site location in proposing a new phylogenetic hypothesis for the species of *Apis*. When this character is considered in conjunction with 20 others, the analysis does not support the hypothesis that the extant cavity-nesting species of honey bees are the basal lineage(s) within *Apis*. However, the phylogenetic hypothesis resulting from this analysis does not unequivocally indicate whether the ancestral *Apis* nested in cavities. If nesting in cavities is taken to be the ancestral condition for the Apidae as a whole, then it is equally parsimonious to argue either that the ancestral *Apis* nested in cavities and *andreniformis* + *florea* and the *dorsata* group have independently moved to nesting in more open conditions (Figure 1.1A, Character 20) or that the ancestral *Apis* evolved the novel behavior of nesting in the open, and the common ancestor of *mellifera*, *cerana*, and *koschevnikovi* subsequently reverted to nesting in cavities (Figure 1.1B). The uncertainty is a consequence of the topology of the cladogram.

Thus, even though the cladistic analysis presented here supports the genealogical relationships among *Apis* species that Lindauer derived from his comparative study of the dance language, that does not mean that every element of his elaborate evolutionary scenario receives unequivocal support. A cladogram is only a hypothesis about common

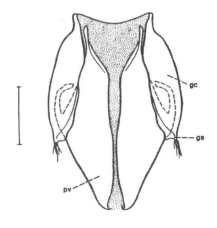

Figure 1.5 Dorsal view of genital capsule (excluding endophallus) of male *Apis dorsata* from the Malay Peninsula; Abbreviations as in Figure 1.3 A.

ancestor relationships. One can use cladograms and parsimony arguments to test hypotheses about sequences of character evolution, because it is possible to determine whether a postulated sequence is the most parsimonious arrangement on a given cladogram. On the other hand, more than one sequence may be equally parsimonious, depending on the tree topology. Hypotheses about the causal factors that may have been responsible for creating the changes in character states are even more indirectly related to the cladogram. Cladistic analyses provide an indispensable historical frame of reference for evaluating hypotheses about evolutionary change, but they cannot and should not be expected to resolve every question or competing theory about evolutionary processes. The analysis presented here does not directly contradict any of Lindauer's ideas about the evolution of the dance language. However, it is best to keep in mind that other possible explanations could be equally consistent with our current hypothesis of phylogenetic relationships among the species of *Apis*. In particular, it is worth considering that the architectural context and behavioral components of the ancestral dance language may not have been retained in unmodified form in any of the extant species of honey bees.

Contribution Number 3043 from the Department of Entomology, University of Kansas, Lawrence, Kansas 66045, USA

Appendix 1

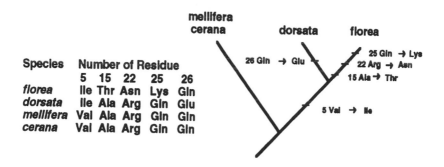

Species	Number of Residue				
	5	15	22	25	26
florea	Ile	Thr	Asn	Lys	Gln
dorsata	Ile	Ala	Arg	Gln	Glu
mellifera	Val	Ala	Arg	Gln	Gln
cerana	Val	Ala	Arg	Gln	Gln

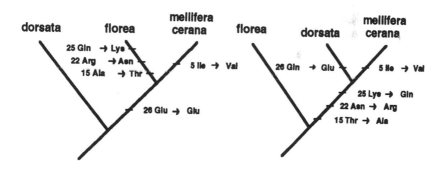

Kreil's (1973, 1975) data on amino acid sequences in the peptide melittin can be used to distinguish three groups in the genus *Apis*: *florea*, *dorsata*, and *mellifera* + *cerana*. (Kreil was working under the assumption that there are only four species in *Apis*. The amino acid sequence is identical in the melittin of *mellifera* and *cerana*.) For any group of three taxa, there are three possible cladograms. In the absence of information on character polarity, Kreil's data can be explained equally well by any of the three possible cladograms. The data matrix used here is based on Table 2 of Kreil (1975), except that residue No. 10 is omitted. This is because each of the three terminal taxa has a different residue at that position, so that it cannot provide any information about common ancestry for the three taxa. On the cladograms, the following notation is used to indicate the postulated character changes: Arabic numerals (e.g., 5, 15) designate the position of the amino residue in the linear sequence within the melittin molecule, the three-letter symbols (e.g., Thr, Ala) are standard abbreviations for amino acid residues (as presented in Kreil, 1975), and the arrows indicate the hypothesized direction of character state transformation (e.g., from Thr to Ala, or from Ala to Thr). All three cladograms are equally parsimonious in the sense that each requires five transformations. Only by establishing which amino acid residue at a given position is plesiomorphic will it be possible to determine which of these equally parsimonious cladograms reflects the actual phylogenetic relationships.

Literature Cited

Alexander, B. 1991. A phylogenetic analysis of the genus *Apis* (Hymenoptera: Apidae). *Ann. Entomol. Soc. Am.* 84:137-149.

Buttel-Reepen, H. von. 1906. Beiträge zur Systematik, Biologie, sowie zur geschichtlichen und geographischen Verbreitung der Honigbiene (*Apis mellifica* L.), ihrer Varietäten under der übrigen *Apis*-Arten. *Mitteilungen aus dem Zoologischen Museum in Berlin* 3 (2):121-201.

Cumber, R. A. 1949. Humble-bees and commensals found within a thirty mile radius of London. *Proc. Royal Entomol. Soc. Lond. (A)* 24:119-127.

Dyer, F. C. 1985. Mechanisms of dance orientation in the Asian honeybee *Apis florea*. *J. Comp. Physiol. A* 157:183-198.

------ 1987. New perspectives on the dance orientation of the Asian honey bees. In *Neurobiology and behavior of honey bees.*, R. Menzel and A. Mercer, eds. Springer-Verlag: Berlin. Pp. 54-65.

Farris, J. S. 1983. The logical basis of phylogenetic analysis. In *Advances in cladistics II.* V. Funk and N. Platnick, eds. Columbia University Press: New York. Pp. 7-36.

------ 1988. Hennig86 Reference, Version 1.5. Port Jefferson Station, New York.

Felsenstein, J. 1983. Parsimony in systematics: biological and statistical issues. *Annu. Rev. of Ecol. Syst.* 14:313-333.

Frisch, K. von. 1967. *The dance language and orientation of bees.* Harvard University Press: Cambridge, MA.

Gerstäcker, C. E. A. 1862. Über die geographische Verbreitung und die Abänderungen der Honigbiene nebst Bemerkungen über die ausländischen Honigbienen der alten Welt. *Festschrift, XI Wander-Versammlung deutscher Bienenwirte.* Potsdam, Kraemer, 75 pp. English translation: 1863. *Annals and Magazine of Natural History, London (3)* 11:270-283, 333-347. Republished in German in Buttel-Reepen, 1906.

Hennig, W. 1950. *Grundzüge einer Theorie der phylogenetischen Systematik.* Deutscher Zentralerlag: Berlin.

------ 1966. *Phylogenetic systematics.* University of Illinois Press: Urbana, IL.

Horn, E. 1975. Mechanisms of gravity processing by leg and abdominal gravity receptors in bees. *J. Insect Physiol.* 21:673-679.

Jander, R. and U. Jander. 1970. Über die Phylogenie der Geotaxis innerhalb der Bienen (Apoidea). *Z. vgl. Physiol.* 66:355-368.

Kimsey, S. L. 1984. A re-evaluation of the phylogenetic relationships in the Apidae. *Syst. Entomol.* 9:435-441.

Kluge, A. G. and J. S. Farris. 1969. Quantitative phyletics and the evolution of anurans. *Syst. Zool.* 18:1-32.

Koeniger, N. 1976. Neue Aspekte der Phylogenie innerhalb der Gattung *Apis*. *Apidologie* 7:357-366.

Kreil, G. 1973. Structure of melittin isolated from two species of honeybees. *FEBS Lett.* 33:241-244.

------. 1975. The structure of *Apis dorsata* melittin: phylogenetic relationships between honeybees as deduced from sequence data. *FEBS Lett.* 54:100-102.

Latreille, P. A. 1804. Notice des Espèces d'Abeilles vivant en grande Société, ou Abeilles proprement dites, et Description d'Espèces nouvelles. *Annales du Musée d'Histoire Naturelle* 5:161-178.

Lepeletier, A. L. M. 1836. *Histoire naturelle des Insectes. Suites à Buffon. Hyménoptères.* Paris, Roret. Vol. I, 547 pp.

Lindauer, M. 1956. Über die Verstandigung bei indischen Bienen. *Z. vgl. Physiol.* 38:521-557.

------. 1961. *Communication among social bees.* Harvard University Press: Cambridge, MA.

Maa, T. 1953. An inquiry into the systematics of the tribus Apidini or honeybees (Hym.). *Treubia* 21:525-640.

Maddison, W. P., M. J. Donoghue and D. R. Maddison. 1984. Outgroup analysis and parsimony. *Syst. Zool.* 33: 83-103.

Mason, W. R. M. 1986. Standard drawing conventions and definitions for venational and other features of wings of Hymenoptera. *Proc. Entomol. Soc. Wash.* 88:1-7.

Mathew, S. and K. Mathew. 1988. The "red" bees of Sabah. *Newsletter of Beekeepers in Tropical and Subtropical Countries* 12, International Bee Research Association, p. 10.

Mayr, E. 1942. *Systematics and the origin of species.* Columbia University Press: New York.

Michener, C. D. 1974. *The social behavior of the bees.* Harvard University Press: Cambridge, MA.

Michener, C. D. 1990. Classification of the Apidae (Hymenoptera). *Univ. Kans. Sci. Bull.* 54:75-164.

Mickevich, M. 1982. Transformation series analysis. *Syst. Zool.* 31:461-478.

Peters, W. C. H. 1862. Naturwissenschaftliche Reise nach   Mossambique auf Befehl seiner Majestät des Königs Friedrich Wilhelm IV in den Jahren 1842 bis 1848 ausgeführt. Druck und verlag von Georg Reimer. (pp. 439-443 contain a discussion of 4 *Apis* species by A. Gerstäcker).

Plant, J. D. and H. F. Paulus. 1987. Comparative morphology of the postmentum of bees (Hymenoptera: Apoidea) with special remarks on the evolution of the lorum. *Z. zool. Syst. Evolutionsforsch.* 25:81-103.

Rinderer, T. E., N. Koeniger, S. Tingek, M. Mardan and G. Koeniger. 1989. A morphological comparison of the cavity dwelling honey bees of Borneo, *Apis koschevnikovi* (Buttel-Reepen, 1906) and *Apis cerana* (Fabricius, 1793). *Apidologie* 20:405-411.

Ruttner, F. 1988. *Biogeography and taxonomy of honey bees.* Springer-Verlag: Berlin.

Ruttner, F., D. Kauhausen, and N. Koeniger. 1989. Position of the red honey bee, *Apis koschevnikovi* (Buttel-Reepen 1906), within the genus *Apis. Apidologie* 20:395-404.

Sakagami, S. F. and C. D. Michener. 1987. Tribes of the Xylocopinae and origin of the Apidae (Hymenoptera: Apoidea). *Ann. Entomol. Soc. Am.* 80:439-450.

Sakai, S., H. Hoshiba and E. Hoshiba. 1986. Integrated taxonomic information on honey bees. *Proc. 30th Int. Congr. Apiculture, Nagoya, Japan.* Apimondia (1986):134-139.

Seeley, T. D. 1985. *Honeybee ecology.* Princeton University Press: Princeton, NJ.

Smith, E. L. 1970. Evolutionary morphology of the external insect genitalia. 2. Hymenoptera. *Ann. Entomol. Soc. Am.* 63:1-27.

Smith, F. 1865. On the species and varieties of the honey-bees belonging to the genus *Apis. Annals and Magazine of Natural History, London (3)* 15:372-380.

Snodgrass, R. E. 1956. The anatomy of the honey bee. Cornell University Press: Ithaca, NY.

Snodgrass, R. E. 1941. The male genitalia of Hymenoptera. *Smithson. Misc. Collect.* 99 (14):1-86, Pl. 1-33.

Sober, E. 1983. Parsimony in systematics: philosophical issues. *Annu. Rev. Ecol. Syst.* 14:335-357.

Tingek, S., M. Mardan, T. E. Rinderer, N. Koeniger and G. Koeniger. 1988. Rediscovery of *Apis vechti* (Maa, 1953): the Saban honey bee. *Apidologie* 19: 97-102.

Winston, M. L. 1987. *The biology of the honey bee*. Harvard University Press: Cambridge, MA.

Winston, M. L. and C. D. Michener. 1977. Dual origin of highly social behavior among bees. *Proc. Natl. Acad. Sci. USA* 74: 1135-1137.

Wongsiri, S., K. Limbipichai, P. Tangkanasing, M. Mardan, T. Rinderer, H. A. Sylvester, G. Koeniger and G. Otis. 1990. Evidence of reproductive isolations confirms that *Apis andreniformis* (Smith, 1858) is a separate species from sympatric *Apis florea* (Fabricius, 1787). *Apidologie* 21:47-52.

Wu, Yan-ru and B. Kuang. 1986. A study of the genus *Micrapis* (Apidae). *Zool. Res.* 7:99-102. [In Chinese.]

------. 1987. Two species of small honeybee--a study of the genus *Micrapis*. *Bee World* 68:153-154. [English translation of Wu & Kuang, 1986.]

2

A Review of the Diversity
of Species Within *Apis*

Gard W. Otis

The genus *Apis* has recently been recognized as being more diverse than was previously believed. Most authors in recent years have recognized only four species (e.g., Free, 1982; Goetze, 1964; Gould and Gould, 1988; Koeniger, 1976; Morse and Hopper, 1985; Ruttner, 1968, 1975; Wilson, 1971). The giant or rock honey bee, *A. dorsata*, and the dwarf honey bee, *A. florea*, are open-nesting species that build a single, exposed comb. The western honey bee, *A. mellifera*, and the eastern honey bee, *A. cerana*, nest in cavities where they construct multiple, parallel combs. Ruttner (1988) has summarized much of what we know about these four traditional species. Recent investigations in Southeast Asia have confirmed the existance of three additional species and several other potential species (Figure 2.1). This chapter will review some of the recent studies that have been undertaken of these new taxa.

Historical Overview

The taxonomy of the honey bees has been confusing, to say the least, for several reasons. First, the geographic variation found in the best known species *Apis mellifera*, is extensive. Some of the numerous races (subspecies) are distinctive enough to have been considered to be species in the past. While the differences between some of these races may be considerable, close examination has demonstrated that the races involved have non-overlapping geographic distributions. There are no documented

29

reproductive barriers between these races, and in several cases the absence of such reproductive barriers has been demonstrated (e.g., between European races and the African *A. m. scutellata*; Kerr and Bueno, 1970). Extensive geographic variation is also known in *Apis cerana* (Ruttner, 1988; Peng et al., 1989 and references therein). Consequently, these distinct morphs must be considered as only subspecies until such time as more detailed studies demonstrate the presence of reproductive isolating mechanisms between some of the parapatric races.

Second, most taxonomic studies of honey bees have been conducted on worker bees. The drones would be more appropriate to investigate but the genitalia pose some problems for study. Unlike most insects, the external genitalia are of little diagnostic use for species-level systematic work. The endophalli are much more distinctive (Simpson, 1960, 1970; Ruttner, 1988), but have been little studied until recently because males are poorly represented in collections and the male endophallus is rarely everted in specimens. A recent study of the morphologies of the uneverted endophalli of *Apis* species may overcome this latter limitation (Koeniger et al., in prep.).

Third, in spite of their much greater biological diversity, the honey bees of Asia have been relatively poorly studied. In part this reflects the paucity of researchers in Asia working on honey bees until recently, but in addition few western researchers have taken the opportunity to work in Asia. For those few who have, the insights they have provided on the bees in that region have given us a tantalizing picture of the differences between the species (c.f., Lindauer, 1956, 1957; Morse and Laigo, 1969; Seeley et al., 1982; Dyer, 1985a, 1985b). Many basic biological facts that are well known for *A. mellifera* of Europe are still unverified for *A. cerana*, the best studied of the Asian honey bees. We have only fragmentary knowledge of the biologies of the giant and dwarf bees.

The extensive geographic variation exhibited by the honey bees has led to a vast number of specific and subspecific names. Early researchers working primarily from museum specimens described forms which differed only slightly from other named forms. Descriptions were often vague and sometimes bees from the same location were described repeatedly as different species or subspecies. The result was "about 600 names which were used at some time in the past for a member of the genus *Apis*" (Ruttner, 1988, p. 59). There have been several attempts to clarify the classification of the genus (e.g., Ashmead, 1904; Buttel-Reepen, 1906; Skorikov 1929) which have helped to eliminate some of that confusion.

One of the last major taxonomic revisions of the honey bees was published in 1953 by T. Maa. He followed Ashmead (1904) in dividing

Figure 2.1 All currently recognized species or possible species in the genus *Apis*. Left to right, top row: *Apis mellifera* (Canada), *A. koschevnikovi* (Sabah, Borneo), *A. cerana* (India), *A. cerana* (Malay Peninsula), *A. florea* (S. India), *A. andreniformis* (Malay Peninsula); bottom row: *A. dorsata* (Malay Peninsula), *A. binghami* (S. Sulawesi), *A. breviligula* (Luzon, Philippines), *A. laboriosa* (northeastern India).

Apis into three genera corresponding to the giant bees, dwarf bees, and cavity-nesting bees, and further recognized two subgenera (relating to *A. mellifera* and *A. cerana*) and a total of 24 "species," primarily on the basis of morphological differences of worker bees. His work was based on verbal descriptions and some diagrams, but lacked a dichotomous key which would allow specimens to be unambiguously identified. In addition, several of the "species" he recognized from different parts of Europe and North Africa, such as *Apis adansonii* and *A. lamarckii*, were even at that time known to constitute subspecies rather than true biological species. Consequently, Maa's work was largely disregarded by most subsequent apicultural researchers. Maa performed a valuable service by summarizing what he recognized as distinct forms and by further clarifying many of the remaining nomenclatural synonomies within the genus.

Without there being any clearer evidence of additional species within *Apis*, most researchers adopted the conservative, four-species view of the genus restated by Goetze (1964). This carried over even into Ruttner's (1988) recent detailed summary of the genus. There have been hints that some of the forms that Maa recognized may in fact represent additional species. Michener (1974) indicated that *A. florea* may be divisible into two species (*florea* and *"andrenoides"*), and that "possibly also *A. dorsata* and *A. cerana* may be divisible into two or possibly more distinct species" (p. 348). Research during the 1980's has clearly established the truth of this cautious statement and has sent researchers scurrying to read Maa's tome. The rest of this chapter will review the exciting developments in *Apis* taxonomy over the past decade.

The *Apis dorsata* Complex

Maa recognized four giant honey bee species in the genus *Megapis* (Figure 2.1). The most widespread of these is *Apis dorsata* F. 1793. It is found throughout India and Southeast Asia, including Palawan, Borneo, and the string of islands of Indonesia from Sumatra to Timor and eastward to the Kai Islands. Because of its large colony size, habit of open-nesting, and well-recognized defensive behavior, *dorsata* has been relatively poorly studied. Most of the literature has been summarized by Ruttner (1988), although some of that information relates to the additional taxa of giant bees that will be discussed below.

Several recent studies have extended our knowledge of *Apis dorsata*. Dyer (1985a) has investigated the communication dances of *dorsata* in detail, including their nocturnal dances. Together with Seeley, he discovered that the the open-nesting species of honey bees (*dorsata, florea*) have flight temperatures well below predicted values which may be indicative of physiological adaptations that have far reaching ecological implications (Dyer and Seeley, 1987). Recent studies on the thermoregulation of both individual bees and colonies of *dorsata* in Malaysia have pointed out some of the problems of heat load that are faced by large, exposed colonies of social insects in the tropics (Mardan, 1989; Mardan and Ashaari, 1990). Large numbers of worker bees fly *en masse* to defecate just prior to the hottest part of the day and/or during windless periods, thereby voiding heat calories and mass which then allows for more efficient cooling (Mardan and Kevan, 1989). In addition, individuals on the external bee-curtain of the colony extrude water droplets on their mouthparts and can alter the tightness of coiling of the aorta in the petiole to regulate their temperature (Mardan and Ashaari, 1990). Studies of marked, newly-emerged workers added to colonies and observed over time demonstrate the existance of well-marked division of labor which differs somewhat from that of *A. mellifera* (Otis et al., 1990). An interesting observation from this latter study was that newly emerged bees and most of the bees in the protective curtain have pale yellow abdomens. In contrast all foragers have orange and black abdomens (Otis et al., 1990). The adaptive significance of this color change is as yet unknown, but may also be integrally related to colony thermoregulation (Mardan and Ashaari, 1990).

In 1980, Sakagami et al. brought attention to "the world's largest honey bee," *Apis laboriosa* F. Smith 1871. Maa (1953) had given a description with key characteristics which distinguish it from other "*Megapis*," including "less prominent ocelli, longer malar areas, much paler thoracic pubescence, and much longer wax-plates." A detailed morphometric analysis using more than 100 characters confirmed the distinctiveness of this bee (Sakagami et al., 1980). *A. laboriosa* nests on cliffs with southerly exposures (Roubik et al., 1985; Underwood, 1986) of the valleys of the Himalaya Mountains, from Nepal east through India, Bhutan, Myanmar (Burma), and China. Active and abandoned nests have been observed at elevations ranging from 1220 m to 3510 m (Underwood, 1990a), and foragers have been caught between 1200-4100 m in Nepal (Roubik et al., 1985) and 930-3450 m in Tibet (Wu, 1982). In contrast, *A. dorsata* rarely nests above 1250 m (Underwood, 1990a).

In Nepal, *A. laboriosa* undergoes an interesting seasonal cycle. Roubik et al. (1985), after reviewing collection data and reports of nesting aggregations, suggested that nests may be occupied all year at protected cliff sites. More detailed observations of nesting sites at different elevations and at different times of year by Underwood (1990a) indicate that no nest sites are occupied throughout the year. At the lowest elevations (below 2000 m) colonies leave the cliffs in Late November and early December. At the same time isolated swarms can be found in inactive clusters on cliffs or tree trunks near the ground. These swarms apparently subsist largely on the contents of their honey stomachs for up to two months in combless clusters, with very little activity except for occasional foraging on warm days. Temperatures of the swarm core can be as low as 6°C, and individual bees can cool to less than 4°C without ill effects (Underwood, 1990b, 1990c). This overwintering strategy is unique for honey bees, although swarms of *A. dorsata* are also reported to occasionally rest for a few weeks without building combs (Lindauer, 1957). After returning to the lower cliff sites in February, some colonies apparently reproduce a few months later, as evidenced by the presence of queen cells along the bottom margin of nests in April and May. As the season progresses, swarms colonize cliffs at higher elevations, in some years reaching the subalpine zone (2800-3600 m) by June or July. These migratory movements may be in part stimulated by absconding following honey harvesting. A reverse migration to lower elevations occurs by October from higher to lower altitudes.

The species status of *Apis laboriosa* has been contested recently. Sakagami et al. (1980) and McEvoy and Underwood (1988) argued that the number of quantitative differences between *dorsata* and *laboriosa*, their sympatric distribution at the upper and lower extremes of their ranges respectively, and the lack of intermediate forms were all evidence that *laboriosa* should be considered a distinct species. However, the countering argument is that this evidence is not sufficient to give this distinct morphotype species status; some of the subspecies of *A. mellifera* are as distinct from each other as are *laboriosa* and *dorsata* (Goetze, 1964; Koeniger, 1976; and Ruttner, 1968, 1988). Further evidence of reproductive isolation is required. Ruttner (1988) suggested that the taxonomic status of *laboriosa* would only be settled when the male endophallus was described, as evidence of a physical barrier to intermating. When the endophallus of *laboriosa* was described shortly thereafter, no species-specific characters were found (McEvoy and Underwood, 1988). (This matter is still unresolved following the redescription of the *dorsata* endophallus which may differ from that of *laboriosa*; Koeniger et al., 1990b). More recently, Underwood (in press 'a') has provided preliminary data on the time of drone flight of *Apis laboriosa*. Whereas *dorsata* drones fly just before dark, *laboriosa* drones fly in early afternoon, from 1220 h to 1420 h. The ocelli of drones, which are noticeably raised in most members of the *dorsata* complex and are undoubtedly an adaptation to flight at the low light intensities encountered at dusk, are flat in *laboriosa* drones (McEvoy and Underwood, 1988; G. W. Otis, pers. obs), which also suggests that mating flights occur during the day and not at dusk. Although additional data would be helpful, this difference in drone flights appears to provide the reproductive isolation needed to consider that *laboriosa* is a valid species.

Underwood (1990c, in press "b") also studied the some metabolic characteristics of individual *laboriosa* workers. His studies indicate that, like *dorsata* and *florea* (Dyer and Seeley, 1991), *laboriosa* is a relatively low powered bee with low wing-loading and relatively slow flight speed. Surprisingly, the body temperature of foragers arriving at feeding dishes was higher when the reward (sugar concentration) was higher. This

Figure 2.2 *Apis koschevnikovi* workers. Note the distended abdomens (nectar storage) and *Varroa* mite on the thorax.

suggests that the bees may vary their flight metabolism according to the anticipated reward, a finding only recently observed in *A. mellifera* (Schmaranzer and Stabentheinher, 1988).

Two other morphotypes of giant honey bees have been recognized by some authors as full species: *Apis binghami* Cockerell 1906 of Sulawesi, nearby Sula Island, and Butung, and *Apis breviligula* Maa 1953 of the Philippines (Luzon, Mindoro, and probably throughout, excluding Palawan). The two are quite similar in color, both being uniformly black with distinct white bands on the abdomen. Both have raised ocelli, as is also the case in *A. dorsata* and are known to forage nocturnally. Some of the differences noted by Maa (1953) are summarized by Ruttner (Table 8.1, 1988). *A. breviligula* is slightly shorter but with broader abdomen and substantially shorter mouthparts than *A. binghami* (Maa, 1953). Nesting aggregations which are common for *dorsata* have not been observed for either *breviligula* (Morse and Laigo, 1969; Starr et al., 1987) or *binghami* (Otis, pers. obs.).

It could be argued that these two morphotypes should be considered separate species because they have isolated populations with distinct morphological features. However, as their distributions are allopatric from that of *A. dorsata*, their status as species is likely to remain an arbitrary decision. Biologically, there should have been no selection for isolating mechanisms, and in fact there is no evidence that premating barriers exist. Both *dorsata* (Koeniger and Wijayagunasekera, 1976; Koeniger et al., 1988) and *binghami* (Otis, pers. obs.) take their mating flights at the same time, after the sun has set in the evening. The genitalia of the *binghami* drone does not apparently differ substantially from that of the *dorsata* male (G. Koeniger, pers. comm.). (The drone flight time and genitalia of *breviligula* have yet to be described.) To demonstrate postmating barriers, one would need to collect semen from drones of one form and inseminate virgin queens of another form. Given the biology of the giant honey bees, this is virtually impossible. In the absence of that type of evidence, one must rely on the results of morphometric analyses (DuPraw, 1965; Sakagami et al., 1980; Ruttner, 1988) or genetic differences (e.g., Chapter 8, this volume). While both of these methods clearly allow for discrimination of species of honey bees, they have also been used extensively for intraspecific racial differentiation of bees of different geographic origin. The magnitude of differences necessary to differentiate species and subspecies (races) is subjective. While these three *dorsata*-type bees can be differentiated morphometrically (Sakagami et al., 1980.) and genetically (D.R. Smith, pers. comm.), other biological criteria are needed to establish unequivocally that they are valid biological species.

Apis koschevnikovi

In 1978, N. Koeniger collected a sample of relatively large, reddish bees in Sumatra (Ruttner et al., 1989). When the morphometric analyses had been completed, it was treated as an aberrant sample of *A. cerana* and was largely forgotten. A decade was to pass before two short but exciting papers were published (Tingek et al., 1988; Koeniger et al., 1988) that recognized that "aberrant sample" for what it was: a distinct species. In their redescription, the authors assigned this bee the name *Apis vechti* Maa 1953. Mathew and Mathew (1988) independently reported on the species from the same experimental station in Sabah, East Malaysia (Borneo).

Figure 2.3 Fully everted genitalia of *A. koschevnikovi* male.

After careful checking, Ruttner et al. (1989) determined that Buttel-Reepen (1906) had described a similar bee collected from "Kamerun" and "North Borneo" (=Sabah) as a distinct species, *Apis koschevnikovi*, a name which clearly has priority. This cavity-nesting bee, now generally referred to as the red honey bee, is substantially larger than the similar *A. cerana* (Rinderer et al., 1989) and is nearly uniformly reddish yellow (Figure 2.2) (Maa, 1953). The cubital index of the worker forewing is large and variable ($\bar{x}$ = 7.46 ± S.D. 3.04). The general coloration is very similar to the cordovan mutation known from *Apis mellifera* (Laidlaw et al., 1953; see Plate 1 of Sakagami et al., 1980).

We presently know almost nothing about the biology of *A. koschevnikovi* except that it is a valid, biological species. Tingek et al. (1988) figured the endophallus which is distinct in many ways from that of *cerana* (Koeniger et al., in prep.), especially the hairy region of the vestibulum and the knobby shape of the upper cornua (Figure 2.3). This by itself is evidence of reproductive isolation from the relatively similar *A. cerana*, but Koeniger et al. (1988) demonstrated that drones took mating flights in late afternoon between 16.45 h and 18.15 h when no drones of sympatric species were flying. Therefore, there is clear reproductive isolation from other *Apis*.

Other information on *A. koschevnikovi* is sparse. Much of what follows is based on the limited experience of M. Mardan and myself and the brief report by Mathew and Mathew (1988), all from Tenom, Sabah, Borneo. Worker bee flight at flowers is very fast and more erratic than that of *cerana*. Colonies of *koschevnikovi* tend to be smaller than those of *cerana*. Bees fly rapidly into their colonies, with no noticeable activity on the outside of the nest. When disturbed, the worker bees hiss in a manner similar to that of *cerana* (Schneider and Kloft, 1971; Koeniger and Fuchs, 1975). Generally workers retreat into the nest cavity upon disturbance, but repeated disturbance will cause a few bees to leave and sting. They do not follow any great distance and calm down quickly after disturbance. *Varroa* mites, apparently *V. jacobsoni* (Figure 2.2) (M. Delfinado-Baker, pers. comm.) but larger than those on sympatric *A. cerana*, infest only the drone cells, which have a pore in the capping similar to that of *cerana*. Queen cells are constructed in moderate numbers along the margins of combs during swarming. Queen tooting (piping) by an emerged queen following swarming consists of several broken tones followed by rapid abdominal thumping of the comb. This is different from reports of *mellifera* queen piping (Wenner, 1962), but is similar to that of *cerana* (Otis and Patton, in prep.). Colonies seem to swarm at small colony size, and abscond frequently (Mathew and Mathew, 1988). One swarm had only 3380 bees. When brood was exchanged between *cerana* and *koschevnikovi* colonies, the *koschevnikovi* colony destroyed the foreign brood, but *koschevnikovi* brood in a *cerana* colony emerged and the bees were accepted into the colony (Mathew and Mathew, 1988). The ecology of *koschevnikovi* is completely unknown.

A. *koschevnikovi* is known to occur only on the islands of Borneo and Sumatra. Specimens in the British Museum indicate that it is widespread over the island of Borneo (Otis, pers. obs.). During my visit to Sabah, I had the impression that this species is associated with forested lands while *cerana* is more common in disturbed habitats. There are very few collections from Sumatra. At present it is known from only three localities, all at mid-elevations: Mt. Tanggamus, 550 m, Lampung Province (Maa, 1953, described as *Apis lieftincki*, with some morphological differences from Bornean specimens given), Muara, near Solok, Sumatera Barat (Ruttner et al., 1989), and Gunung Leuser National Park, Aceh Province (specimens in the Royal Ontario Museum, G. Otis, pers. obs.).

Figure 2.4 Foraging worker of *Apis andreniformis*.

Apis andreniformis

This dwarf honey bee was originally described from collections made by Alfred Russell Wallace in Borneo (F. Smith, 1858), but was subsumed with all dwarf bees under the name *A. florea* shortly thereafter (Smith, 1865). Maa (1953) also recognized two species of *Micrapis*. However, it was only a few years ago that two separate dwarf bee species were clearly differentiated (Wu and Kuang, 1986, 1987). Key characteristics for workers include the color of hairs on the hind tibia and dorsolateral surface of the hind basitarsis (black in *andreniformis*, white in *florea*) and proportions of the malar space (wider than long in *andreniformis*, longer than wide in *florea*) (Wu and Kuang, 1986, 1987). The cubital index differs as well, but with considerable overlap in values for individual bees (*andreniformis*: $\bar{x}$ = 6.07, range 3.5 to 13.8; *florea*: $\bar{x}$ = 2.78, range = 1.93 to 5.27) (Wongsiri et al., 1990). Standard multivariate analyses of morphometric measurements separate *andreniformis* and *florea* into two adjacent and non-overlapping clusters of points (Ruttner, pers. comm.). Color is usually distinctive, with typical *florea* workers having the first two and part of the third terga reddish brown while *andreniformis* are often all

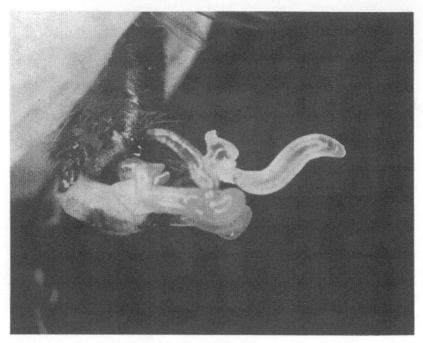

Figure 2.5 Fully everted genitalia of *A. andreniformis* male.

black with three distinctive white bands (Figure 2.4). However, there is
considerable variation in color, with some *andreniformis* having completely
reddish-brown abdomens (possibly young bees) or in some cases being
almost indistinguishable from *florea*. All three color types and inter-
mediates can sometimes be collected from the same colony. (It was on the
basis of this color variation that Smith [1865] retracted his designation of
andreniformis as a species.) Drones of both species have a lobe on the
hind tibia, which is less than half the length of the tibia in *andreniformis*
and more than half the length of the tibia in *florea* (Wu and Kuang, 1987;
Wongsiri et al., 1990).

The male genitalia of *florea* and *andreniformis* are clearly more similar
to each other than to any other *Apis* species (Koeniger et al., in prep.).
Moreover, they have several species-specific differences (Wongsiri et al.,
1990; Koeniger et al., in prep.), such as the shape of the terminal portion
of the endophallus, which are *si paret* evidence of reproductive isolation
between the species (Figure 2.5). The shape of the endophallus, ending in
a thin tip, suggests that there may be direct sperm transfer in *andreniformis*
(Koeniger et al., 1990a) as has been reported for *A. florea* (Koeniger et al.,

Figure 2.6 Typical small colony of *A. andreniformis*. Nest defenders are beginning to accumulate at the bottom of the colony.

1989). *A. andreniformis* queens multiply mate with at least 7-8 drones, as evidenced by the number of sperm in the spermatheca of two mated queens (0.98 and 1.09 million) as compared to the number of sperm in the seminal vesicles of males ($\bar{x} = 130,000 \pm 10,000$, n = 5) (Koeniger et al., 1990a).

At present there are no data concerning reproductive isolation on the basis of mating flight times. Preliminary observations for *florea* drones in Bangkok (S. Wongsiri, unpublished) and *andreniformis* drones in Peninsular Malaysia (M. Mardan and G. Otis, unpublished) indicate no substantial differences in mating flights, with a peak of male activity at approximately 13.00 h. However, in these particular locations there is only one species of dwarf honey bee present, so there could have been no directional selection for different times of mating activity. Additional data are required from sites where these two species are sympatric or parapatric.

The biology of *Apis andreniformis* is currently virtually unknown. The information that follows is based on the experiences of S. Wongsiri in Thailand and M. Mardan in Peninsular Malaysia, as well as my experience with the species from Peninsular Malaysia and Sabah, East Malaysia (Otis et al., in prep. 'b'). Colonies are small (Figure 2.6) and generally well concealed beneath vegetation from ground level to heights of 5 m (the local name in southern Malaysia is "nyepan," meaning "hidden"). The cells in the small honey storage area and dance platform are oriented slightly below horizontal, in contrast to *florea* combs where some cells point upward. Mature colonies in Malaysia had populations of 4150 ± 917 bees (n = 4) and an average comb area of only 167 ± 40.2 cm^2 (n = 4), which is smaller than reported from southern China (Wu and Kuang, 1987). Initial attempts to capture entire colonies resulted in many bees following for some distance and stinging (see also Wu and Kuang, 1987). Bees prepared to defend their colony by moving to the lower margin of the nest where they formed a chain of active bees. To capture colonies during daylight, it was necessary to heavily spray the bees on the nests with water. The bees responded by moving upward onto the upper flanks of the nest, in effect creating a living umbrella of bees, as is also true of *florea*. They also began to move their wings and flick the water off almost immediately, so the calming effects of the spray were short-lived. In contrast to what was reported by Otis (1990), a previously undescribed mite, *Euvarroa wongsirii* (Lekprayoon and Tangkanasing, in press; Otis et al., in prep 'a'), has been found infesting drone brood. Colonies are occasionally attacked by honey buzzards (*Pernis apivorus*), but are relatively well protected from ants by rings of propolis placed by the bees on the comb-supporting branches as is also done by *florea*. The honey is used to treat specific ailments by rural healers in the Malay Peninsula (Ahmad Razuli, pers. comm.). This species was abundant in southern Malaysia in coconut growing areas near sea-level, common at the Tenom Research Station in Sabah, but relatively scarce in the vicinity of Kuala Lumpur. It is scarcer than *florea* in both Thailand (Wongsiri, pers. comm.) and southern China (Li et al., 1986).

The distributions of *andreniformis* and *florea* have now been determined (Otis et al., in prep. 'b'). The western portion of the range of *A. florea* as depicted by Ruttner (1988) is correct up to the Thai-Malaysia border. From there southward and eastward on the islands of Sumatra, Java, Borneo, and Palawan, *andreniformis* is the only species of dwarf bee except in lowland Java (e.g., Jakarta) where *florea* has a disjunct population. Both species are found throughout Indochina, and *andreniformis* has been collected as far west as the Shillong Hills and near Darjeeling, India. In the allopatric portions of their ranges (e.g., *florea* in southern India and *andreniformis* in the Malay Peninsula and Borneo), both species occur from sea level to 1400 m, with the bees from higher elevations being substantially larger in size. Where the two species are found in the same geographic area (Java: Maa, 1953; northern and eastern Thailand and northeastern India: Otis et al., in prep. 'b'), *florea* is the lowland species while *andreniformis* is found at higher elevations. This contrasts with information from China which reports *florea* as occurring up to 1900 m, while *andreniformis* is found below 1000 m (Wu and Kuang, 1986, 1987).

Conclusion

It has become obvious that the diversity of honey bees in Asia is much greater than has been recognized in recent years. Particularly in Indonesia, the Philippines, and Malaysia, the numerous islands are inhabited by disjunct populations of bees, geographically isolated from conspecifics by water gaps. In such a setting one would expect to find extensive variation in the bees from different islands, in some cases having resulted in speciation which is common in island settings. It is surprising that so little attention has been directed to the bees of this region. Ruttner (1988), after correctly pointing out that *A. cerana* from Borneo and Sulawesi and the eastern part of Indonesia had not been studied morphometrically, stated that "it is not very likely that a completely different major type [of bee] remained undetected" (p. 156). The gross underestimation of this statement became strikingly clear less than a year later with the rediscovery of *Apis koschevnikovi* on Borneo. Further collecting and study in this region should resolve most remaining uncertainties surrounding the taxonomy of the honey bees.

Two situations seem particularly interesting and deserving of further study. I was particularly struck by the appearance of two small, black, and short-winged giant honey bees in the collections of the British Museum, with collection locality of Port Blair, Andaman Islands, India. When I visited Port Blair in 1990, I was disappointed to find *Apis dorsata* that differed little in external appearance from *dorsata* in southern India. It is possible that the specimens in the British Museum were actually collected in the neighboring and difficult-to-visit Nicobar Islands. The bees on both of these groups of isolated oceanic islands need to be more fully surveyed.

The second situation involves the *cerana*-like bees of Sulawesi. In South Sulawesi there are two very different populations of *Apis cerana*.

The coastal form is very small, black, and extremely defensive, whereas the interior form from the higher and more forested regions is at least 50% larger, yellow, and relatively gentle. In one location, near Bulukumba, the two forms can be found within 12 km of each other. These could represent highly evolved ecotypes, but further surveys are required, especially to determine whether the two forms occur sympatrically in some locations and whether there is any reproductive isolation by time of drone flight or by other means. Preliminary observations of the male genilatia have not indicated any differences between the two morphotypes.

This review has touched on several studies to be undertaken in the future, the results of which would prove to be very interesting. Some of these are:

1. quantification of the mating flight times of *florea* and *andreniformis* drones from sites where they occur allopatrically and sympatrically;
2. the ecologies of *florea* and *andreniformis* in both allopatric and sympatric settings to understand what factors result in their altitudinal separation where they are sympatric, and their geographical separation near the Thai-Malaysia border;
3. the physiology of *Apis laboriosa* swarms in their winter clusters which enables the bees to survive for two months at such low body temperatures;
4. the adaptive value of the color change in *dorsata* as the bees shift from curtain bees to foragers (*andreniformis* may undergo a similar change in color);
5. the comparative ecologies of the Asian honey bees, where there are frequently four sympatric species (e.g., E. India, N. Thailand, Java, where there are two dwarf bee species; Borneo and Sumatra where there are two cavity-nesting species);
6. further survey of the Indonesian and Philippine archipelagos, where there has been virtually no collecting on many islands. Some forms that Maa (1953) recognized deserve further attention, such as *Apis peroni* from Timor, *A. samarensis* of Samar, the bees south to Mindanao in the Philippines (not mentioned by Maa), and *A. johni* of Sumatra. Of this last form, Maa wrote, "From the original description, this remarkable species is the most primitive member of the genus" and has several unique features, including smoky wings.
7. the timing and events that are most likely to have contributed to speciation of honey bees.

In addition, many of the discoveries concerning honey bees that have been made by studying *Apis mellifera* need to be verified in other honey bee species in order to determine if they are general characteristics of *Apis* or specific adaptations of *A. mellifera*.

The honey bees, genus *Apis*, consist of a highly evolved, behaviorally complex group of insects. From this review of our current knowledge of the species of honey bees, it is obvious that our understanding of the genus *Apis* is far more rudimentary than we previously thought. The recent recognition of at least seven and possibly more species presents incredible opportunities for apicultural researchers to make important, novel discoveries about honey bees. Those who take up this challenge are likely to contribute in important ways to our understanding of diversity of honey bees.

Acknowledgments

This chapter is an outgrowth of an interest in the bees of Asia sparked by Wasantha Punchihcwa of thc Ministry of Agriculturc, Sri Lanka, Makhdzir Mardan of the Universiti Pertanian Malaysia, and a consultancy visit to the Malaysia Beekeeping Project, funded by the International Development Research Centre of Canada. During my sabbatical leave in 1989, made possible by a Research Leave Grant from the University of Guelph and travel funds from IDRC, many people facilitated my travel and bee collections. These individuals were Makhdzir Mardan, Norazhar Zainal, Nordin Sarini, Fauzi, and Mohammed Muid at UPM; H.T. Tang, Kenny Chiew Kean Yong, Yusrin Yusof, and Clive Marsh of Yayasan Sabah; Tay Oeng Boek, Mathew Tulas, Salim Tingek, Jun Athanasius, and K.P and Susan Mathew of the Dept. of Agriculture, Sabah (most at the Agricultural Research Station, Tenom, Sabah); Peter Keating in the Philippines; Soesilawati Hadisoesilo, Forestry Department, Bangkinang, Sumatra; Bakir Ginoga, Busra Kia, and Iwan Setiawan of the Department of Forestry, Raja Andi and Mogana Abbas of B.R.L.K.T., and Iyus Ramelan of Balai Penelitian Kehutanan, South Sulawesi, Indonesia; and Pongthep Akratanakul and Siriwat Wongsiri of Bangkok, Thailand. Of course, beekeepers and honey hunters too numerous to mention were helpful in allowing me to observe their activities and sample their bees wherever I went in Asia.

Literature Cited

Ashmead, W. H. 1904. Remarks on honey bees. *Proc. Entomol. Soc. Wash.* 6:120-123.

Buttel-Reepen, H. 1906. Apistica. Beiträge zur Systematik, Biologie, sowie zur geschichtlichen und geographischen Verbreitung der Honigbiene (*Apis mellifica* L.), ihrer Varietäten und der übrigen *Apis*-Arten. *Mitt. Zool. Museum Berlin* 3:117-201.

DuPraw, E. 1965. Non Linnean taxonomy and the systematics of honeybees. *Syst. Zool.* 14:1-24.

Dyer, F. C. 1985a. Nocturnal orientation by the Asia honey bee, *Apis dorsata. Anim. Behav.* 33:769-774.

------. 1985b. Mechanisms of dance orientation in the Asian honey bee *Apis florea. J. Comp. Physiol.* A 57:183-198.

Dyer, F. C. and T. D. Seeley. 1987. Interspecific comparisons of endothermy in honey-bees (*Apis*): Deviations from the expected size-related patterns. *J. Exp. Biol.* 127:1-26.

------. 1991. Nesting behavior and the evolution of worker tempo in four honey bee species. *Ecology* 72:156-170.

Free, J. B. 1982. *Bees and Mankind.* George Allen and Unwin: London.

Goetze, G. K. L. 1964. Die Honigbiene in natürlicher und künstlicher Zuchtauslese. I. Systematik, Beugung und Vererbung. *Z. Angew. Entomol. Beihefte.* Nr. 19. P. Parey: Hamburg.

Gould, J. L. and C. G. Gould. 1988. *The Honey Bee.* W. H. Freeman and Co.: New York.

Kerr, W. E. and D. Bueno. 1970. Natural crossing between *Apis mellifera adansonii* and *Apis mellifera ligustica. Evolution* 24:145-148.

Koeniger, G., N. Koeniger, M. Mardan, G. W. Otis, and S. Wongsiri. In prep. Comparative anatomy of male genital organs in the genus *Apis. Apidologie.*

Koeniger, G., N. Koeniger, M. Mardan, R. W. K. Punchihewa, and G. W. Otis. 1990a. Numbers of spermatozoa in queens and drones indicate multiple mating of queens in *Apis andreniformis* and *Apis dorsata. Apidologie* 21:281-286.

Koeniger, G., M. Mardan, and F. Ruttner. 1990b. Male reproductive organs of *Apis dorsata. Apidologie* 21:161-164.

Koeniger, N. 1976. Neue Aspekte der Phylogenie innerhalb der Gattung *Apis. Apidologie* 7:357-366.

Koeniger, N., and S. Fuchs. 1975. Zur Kolonieverteidigung östlicher Honigbienen. *Z. Tierpsychol.* 37:99-106.

Koeniger, N., G. Koeniger, S. Tingek, M. Mardan, and T. E. Rinderer. 1988. Reproductive isolation by different time of drone flight between *Apis cerana* Fabricius, 1793 and *Apis vechti* (Maa, 1953). *Apidologie* 19:103-106.

Koeniger, N., G. Koeniger, S. Wongsiri. 1989. Mating and sperm transfer in *Apis florea. Apidologie* 20:413-418.

Koeniger, N. and H. N. P. Wijayagunasekera. 1976. Time of drone flight in the three Asiatic honeybee species (*Apis cerana, apis florea, Apis dorsata*). *J. Apic. Res.* 15:67-71.

Laidlaw, H. H., M. M. Green and W. E. Kerr. 1953. Genetics of several eye color mutants in the honey bee. *J. Hered.* 44:246-250.

Lekprayoon, C. and P. Tangkanasing. (In press) *Euvarroa wongsirii*, new species of bee mite from Thailand. *Internat. J. Acarol.*

Li, S., Y. Meng, J. T. Chang, J. Li, S. He, and B. Kuang. 1986. A comparative study of esterase isozymes in 6 species of *Apis* and 9 genera of Apoidea. *J. Apic. Res.* 25:129-133.

Lindauer, M. 1956. über die Verständigung bei indischen Bienen. *Z. vgl. Physiol.* 38:521-557.

------. 1957. Communication among the honeybees and stingless bees of India. *Bee World* 38:3-14, 34-39.

Maa, T. 1953. An inquiry into the systematics of the tribus *Apidini* or honeybees (Hym.). *Treubia* 21:525-640.

Mardan, M. 1989. *Thermoregulation in the Asiatic giant honeybee Apis dorsata F. (Hymenoptera: Apidae).* PhD Dissertation. University of Guelph, Guelph, 169 pp.

Mardan, M. and A. H. Ashaari. 1990. Some ecological and physiological factors affecting the thermoregulation of the Asiatic giant honey bee (*A. dorsata* F.). In *Social Insects and the Environment.*, G.K. Veeresh, B. Mallik, C.A. Viraktamath, eds. Oxford and IBH Publ. Co.: New Delhi. Pp. 725-726.

Mardan, M. and P. G. Kevan. 1989. Honeybees and 'yellow rain.' *Nature* 341:191.

Mathew, S. and K. Mathew. 1988. The "red" bees of Sabah. *Newsletter of Beekeepers in Tropical and Subtropical Countries, 12, International Bee Research Association,* p. 10.

McEvoy, M. V. and B. A. Underwood. 1988. The drone and species status of the Himalayan honey bee, *Apis laboriosa* (Hymenoptera: Apidae). *J. Kans. Entomol. Soc.* 61:246-249.

Michener, C. D. 1974. *The Social Behavior of the Bees.* Harvard University Press.: Cambridge, MA.

Morse, R. A. and T. Hopper. (eds.) 1985. *The Illustrated Encyclopedia of Beekeeping.* E.P. Dutton, Inc.: New York.

Morse, R. A. and F. M. Laigo. 1969. *Apis dorsata* in the Philippines. *Monogr. Philippine Assoc. Entomol.* 1:1-96.

Otis, G. W. 1990. Diversity of *Apis* in Southeast Asia. In *Social Insects and the Environment.*, G. K. Veeresh, B. Mallik and C. A. Viraktamath, eds. Oxford and IBH Publ. Co.: New Delhi. Pp. 104-105.

Otis, G. W., M. Mardan and K. McGee. 1990. Age polyethism in *Apis dorsata*. In *Social Insects and the Environment.*, G. K. Veeresh, B. Mallik and C. A. Viraktamath, eds. Oxford and IBH Publ. Co.:New Delhi. P. 378.

Otis, G. W., C. Morin, and E. E. Lindquist. In prep. "a". Notes on the biology of *Euvarroa wongsiri*, a parasite of the dwarf honey bee, *Apis andreniformis. Ann. Entomol. Soc. Am.*

Otis, G. W. and K. Patton. In prep. Queen piping of the cavity-nesting honey bees of Asia. *Apidologie.*

Otis, G. W., S. Wongsiri and M. Mardan. In prep. "b". Notes on the biology and distribution of *Apis andreniformis. J. Kans. Entomol. Soc.*

Peng, Y. S., M. E. Nasr and S. J. Locke. 1989. Geographical races of *Apis cerana* Fabricius in China and their distribution. Review of recent Chinese publications and a preliminary statistical analysis. *Apidologie* 20:9-20.

Rinderer, T. E., N. Koeniger, S. Tingek, M. Mardan, and G. Koeniger. 1989. A morphological comparison of the cavity dwelling honeybees of Borneo *Apis koschevnikovi* (Buttel-Reepen, 1906) and *Apis cerana* (Fabricius, 1793). *Apidologie* 20:405-411.

Roubik, D. W., S. F. Sakagami and I. Kudo. 1985. A note on distribution and nesting of the Himalayan honey bee *Apis laboriosa* Smith (Hymenoptera: Apidae). *J. Kans. Entomol. Soc.* 58:746-749.

Ruttner, F. 1968. Systematique du genre *Apis*. In *Traité de biologie de l'abeille*, Vol. 1. R. Chauvin, ed. Masson: Paris. Pp. 2-26.

------. 1975. Races of bees. In *The Hive and the Honey Bee*. Dadant and sons, eds. Dadant and Sons: Hamilton, IL. Pp. 19-38.

------. 1988. *Biogeography and Taxonomy of Honeybees*. Springer-Verlag: Berlin.

Ruttner, F., D. Kauhausen, and N. Koeniger. 1989. Position of the red honey bee, *Apis koschevnikovi* (Buttel-Reepen 1906), within the genus *Apis*. *Apidoligie* 20:395-404.

Sakagami, S. F., T. Matsumura and K. Ito. 1980. *Apis laboriosa* in Himalaya, the little known world largest honeybee (Hymenoptera, Apidae). *Insecta Matsumurana* 19:47-77.

Schmaranzer, S. and A. Stabentheiner. 1988. Variability of thermal behavior of honeybees on a feeding place. *J. Comp. Physiol. B* 158:135-141.

Schneider, P. and W. Kloft. 1971. Beobachtungen zum Gruppenverteidigungsverhalten der östlichen Honigbiene *Apis cerana* Fabr. *Z. Tierpsychol.* 29:337-342.

Seeley, T. D., R. H. Seeley and P. Akratanakul. 1982. Colony defense strategies of the honeybees in Thailand. *Ecol. Monogr.* 52:43-63.

Simpson, J. 1960. Male genitalia of *Apis* species. *Nature* 185:56.

------. 1970. The male genitalia of *Apis dorsata* (F.) (*Hymenoptera: Apidae*). *Proc. R. Entomol. Soc. Lond. (A)* 45:169-171.

Skorikov, A. S. 1929. Eine neue Basis für eine Revision der Gattung *Apis* L. *Rep. Appl. Ent., Leningrad* 4:249-264.

Smith, F. 1858. Catalogue of the Hymenopterous insects collected at Sarawak, Borneo; Mount Ophir, Malakka; and at Singapore, A. R. Wallace, ed. *J. Proc. Linnean Soc., London Zool.* 2:42-130.

------. 1865. On the species and varieties of the honey-bees belonging to the genus *Apis*. *Ann. & Mag. Nat. Hist.* Ser. 3, Vol. 15:1-9 + Plate XIX.

Starr, C. K., P. J. Schmidt and J. O. Schmidt. 1987. Nest-site preferences of the giant honey bee, *Apis dorsata* (Hymenoptera: Apidae), in Borneo. *Pan-Pac. Entomol.* 63:37-42.

Tingek, S., M. Mardan, T. E. Rinderer, N. Koeniger and G. Koeniger. 1988. Rediscovery of *Apis vechti* (Maa, 1953): The Saban honey bee. *Apidologie* 19:97-102.

Underwood, B. A. 1986. *The natural history of Apis laboriosa Smith in Nepal*. M.Sc. Thesis, Cornell University, Ithaca, NY.

------. 1990a. Seasonal nesting cycle and migration patterns of the Himalayan honey bee *Apis laboriosa*. *National Geographic Research* 6:276-290.

------. 1990b. Bee cool. *Natural History* Dec. 1990:50-57.

------. 1990c. *The behavior and energetics of high-altitude survival by the Himalayan honey bee, Apis laboriosa*. Ph.D. dissertation, Cornell University, Ithaca, NY.

------. In press 'a'. Time of drone flight of *Apis laboriosa* in Nepal. *Apidologie*.

------. In press 'b'. Thermoregulation and energetic decision-making in the honey bees *Apis cerana*, *Apis dorsata*, and *Apis laboriosa*. *J. Exp. Biol.*

Wenner, A. M. 1962. Communication with queen honey bees by substrate sound. *Science* 138:446-447.

Wilson, E. O. 1971. *The Insect Societies*. Harvard University Press: Cambridge, MA.

Wongsiri, S., K. Limbipichai, P. Tangkanasing, M. Mardan, T. Rinderer, H. A. Sylvester, G. Koeniger and G. Otis. 1990. Evidence of reproductive isolation confirms that *Apis andreniformis* (Smith, 1858) is a separate species from sympatric *Apis florea* (Fabricius, 1787). *Apidologie* 21:47-52.

Wu, Y. R. 1982. Hymenoptera: Apoidea. *Insects of Xizang [Tibet]* 2:379-426.

Wu, Y. R. and B. Kuang. 1986. [A study of the genus *Micrapis* (Apidae)]. *Zool. Res.* 7:99-102, *In Chinese*.

Wu, Y. and B. Kuang. 1987. Two species of small honeybee -- A study of the genus *Micrapis*. *Bee World* 68:153-155. (Translation of Wu and Kuang, 1986).

3

Morphological Analysis of the Tribes of Apidae

Michael Prentice

Knowledge of the relationship of *Apis* to the other members of the Apidae is important in a discussion concerning the diversity within the genus. By placing *Apis* in relation to the other Apidae, ideas as to the evolutionary events that led to the origin of the genus and its subsequent diversification may be more critically evaluated. It is not the author's intent, however, to provide a thorough phylogenetic analysis of the relationship of *Apis* to other bees. Such an analysis will be presented elsewhere (Prentice and Daly, in prep.). Instead, this chapter is meant as an outline of some of the major ideas on the relationships within the Apidae and to present some new morphological evidence for a phylogenetic scheme that is contrary to recent ideas concerning those relationships.

The Apidae constitute a family of long-tongued bees consisting of four holophyletic tribes: (1) Euglossini, or orchid bees; (2) Bombini, or bumble bees; (3) Meliponini, or stingless bees; and (4) Apini, or honeybees. Analyses of the relationships within each of these tribes are given by Kimsey (1987) for Euglossini, Williams (1985) for Bombini, Wille (1979a) for Meliponini, and Byron Alexander (Chapter 1, this volume) for Apini. Here we are concerned with the relationships between, rather than within these four tribes.

The Euglossini comprise a group of five genera of strictly Neotropical bees. They are quite specialized as bees in a number of morphological and behavioral characteristics. For instance, they have extremely elongated proboscides -- the longest found in any bee (Figure 3.1E) -- and male

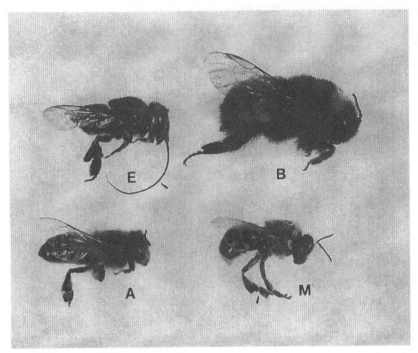

Figure 3.1   The four tribes of Apidae: E = Euglossini (*Euglossa*), B = Bombini (*Bombus*), M = Meliponini (*Melipona*), A = Apini (*Apis*). Note the elongated proboscis of *Euglossa* and the lack of an auricle in *Melipona*.

euglossines are unusual for their highly modified rear tibiae which are used in the collection and carrying of floral fragrances (Kimsey, 1984a), the exact function of which remains obscure (Kimsey, 1987). As apids they are unusual for their lack of eusociality; the genera *Euglossa*, *Eulaema*, and *Eufriesia* consist of solitary to parasocial forms only. The other Euglossine genera, *Exaerete* and *Aglae*, are cleptoparasites of the above genera.

The Bombini (Figure 3.1B) consist of two genera (*Bombus* and *Psithyrus*) of rather large bees which are primarily holarctic in distribution. Non-parasitic species are primitively eusocial having both worker and queen castes in which colonies are founded by lone queens (Michener, 1958). *Psithyrus* and some *Bombus* species are social parasites of social Bombini.

The Meliponini (Figure 3.1M) are a very diverse group of more than 250 distinct species (Kerr and Maule, 1964) that are pantropically distributed with their greatest diversity in the Neotropics. All meliponines

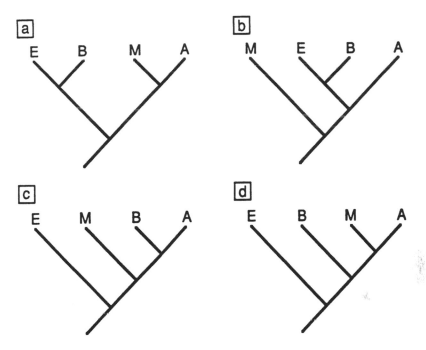

Figure 3.2 Four phylogenetic schemes for the relationships of the tribes of Apidae; diagram (a) = scheme as shown by Michener (1974), (b) = Winston and Michener (1977), (c) = Plant and Paulus (1987), (d) = Michener (1944) and proposed in the present paper. Abbreviations for tribal names are as follows: E = Euglossini, B = Bombini, M = Meliponini, A = Apini.

are highly eusocial bees which live in perennial colonies founded by a queen and a number of workers. They have attained a level of sociality that is, in many respects, equal to that of *Apis* (Michener, 1974). Wille (1979a) recognized eight genera worldwide of which the New World genus *Melipona* is most distinctive.

The Apini (Figure 3.1A) contain the lone genus *Apis* which is native only to the Old World with greatest diversity in tropical Asia. The Apini are less diverse and have far fewer species than the other apid tribes.

While these tribes are easily distinguished on the basis of a number of uniquely derived (autapomorphic) characters, the problem of their inter-relationship has been made difficult because of the great morphological gaps between them. We know, however, that the four tribes taken together form a holophyletic taxon because of a number of shared derived characters (synapomorphies) uniting them (Winston and Michener, 1977; Sakagami and Michener, 1987). For example, apids possess a corbicula on the tibia of the hind leg (Figure 3.4) which, with one exception (the South American anthophorid genus *Canephorula*, which has an analogous structure) is not known in other bees and is therefore a synapomorphy uniting the family.

The Apidae are thought to have evolved from, or at least be closely related to the bee family Anthophoridae (Michener, 1944). Exactly what group of bees within the very diverse Anthophoridae is most closely related to the Apidae, or in other words, forms the sister to the family, is a more difficult question. Sakagami and Michener (1987) supported the view that the Xylocopinae of the Anthophoridae formed the sister to the Apidae based on the similarity between Euglossini and *Manuelia* (of the Xylocopinae) in the form of their seventh and eighth metasomal sterna and because of other similarities between the Apidae and Xylocopinae. They also pointed out that this idea was supported by the fact that direct food transferring between bees is known only from the Apidae and Xylocopinae (Michener, 1972). Therefore, for purposes of phylogenetic reconstruction, the Xylocopinae and other anthophorids should form the main "outgroup" for determining the ancestral and derived character states within the Apidae.

Earlier Ideas on the Relationships of the Tribes of Apidae

Before 1977, the highly eusocial Apini and Meliponini had long been considered to be closely related. For example, Latreille (1802) included the Meliponini in the genus *Apis* along with the honey bees demonstrating that he regarded the Meliponini as highly similar to the Apini. Later classifications elevated the Meliponini to the tribal level, but still classed them closest to *Apis*. For instance, Lepeletiere (1836) classed his Meliponites (= Meliponini) and Apiarites (= Apini) in the same family Apiarides (= Apidae) separate from the other apid tribes. The basis for considering the Meliponini as closely related to the Apini had much to do with the level of sociality in these two tribes; the Meliponini and Apini are the only highly eusocial bees (Michener, 1974). Lepeletiere also classed the Euglossini with the related bee family Anthophoridae, separate from the other apid tribes. This points out that, to the phenetic taxonomists of the last century, the Euglossini and Anthophoridae had been recognized as similar. Although Ashmead (1899) did not explicitly state a phylogeny for the Apidae in his work on the classification of the bees (= superfamily Apoidea), he seems to have supported the relationships as diagrammed in Figure 3.2d. Again, he considered the Meliponini and Apini as most closely related due to their degree of sociality and because of certain mor-

Table 3.1 Morphological characters uniting the Apini with the Meliponini vs. characters uniting the Apini with the Euglossini and Bombini.

Characters uniting Apini with:

A. Meliponini	B. Bombini and Euglossini
1. Prosternum with constriction	1. Auricle present
2. Prosternal apophyseal pit obscured	2. Antenna cleaner
3. Prosternal cuticle smooth, lacking setae	3. Small stigma
4. Hypopharyngeal lobes transversely oriented	4. Basistipital process reduced
5. "Galeal bar" present in maxillus	5. Subgaleal sclerite
6. Stipites not flared above cardines	
7. Antero-lateral process of mesoscutum flat	
8. Metatibial spurs lost	
9. Cuticle with bands of uneven sclerotization	
10. Mandibular grooves lost	

phological features such as the lack of rear tibial spurs (almost all bees, including Bombini and Euglossini have one or two rear tibial spurs) and similarity in forewing venation in these two tribes. Unlike Lepeletiere, however, he recognized that the Euglossini were more closely related to the other apid tribes than to the Anthophoridae. Michener (1944), who began a more modern approach to bee classification using greater numbers of taxonomic characters, diagrammed these same relationships, pointing out other structural similarities between the Meliponini and Apini such as their protuberent scutella and form of epistomal suture. Maa (1953, pg. 628), in his work on the species of *Apis*, also recognized this stating: "Both from morphological and biological evidences, Euglossini is certainly the most primitive tribus of [Apidae], and stands far apart from any of the others. Bombini is the next primitive one and is closer to [Apini] and Meliponini rather than to Euglossini. It appears, however, that there is no intimate relationship to any of the latter three. On the other hand, there exists a close relationship between [Apini] and Meliponini as their lines of development run parallel in many ways..."

Later, however, Michener (e. g., 1964, 1974) diagrammed the relationships for the apid tribes as seen in Figure 3.2a. This differed from his earlier diagram in that the Euglossini and Bombini were shown as sister taxa. That is, the Euglossini + Bombini were considered to be a holophyletic taxon (Bombinae) separate from the holophyletic Meliponini + Apini (Apinae). At that time, however, such a scheme was probably based more on overall phenetic similarities between the Euglossini and Bombini, such as their robust form, than to any cladistic treatment of their character states.

Winston and Michener (1977) changed the long-held view that the Meliponini and Apini were sister taxa by presenting evidence indicating the Meliponini were not, in fact, closely related to the Apini, but instead formed the sister to the rest of the Apidae (Figure 3.2b). This rethinking was largely based on the apparent primitiveness of the Meliponini in two characters of the maxillus which supposedly linked this tribe with the outgroup, Xylocopinae, as well as in their apparently primitive rear basitarsus. They also presented synapomorphies in support of the holophyly of the Euglossini + Bombini. Because of these relationships, and because they considered a loss of highly eusocial behavior in the Bombinae unlikely (which would be necessary in their scheme if highly eusocial behavior in the Meliponini and Apini were homologous) they hypothesized a dual origin of highly eusocial behavior in the Apidae (Figure 3.6b and c). In support of this claim, they emphasized the differences in social systems between the Meliponini and Apini as pointed out by Sakagami (1971). Kimsey (1984b), in a later study of the tribes of Apidae, presented other characters in support of these same relationships, while criticizing the usefulness of some of Winston and Michener's characters. Recent works on bees such as Winston (1987), Ruttner (1988), and Roubik (1989) have accepted Winston and Michener's arguments concerning a dual origin of highly eusocial behavior. Roubik (1989, pg. 296) even went so far as to say: ". . . their independent evolution of permanent sociality appears unquestionable".

Present Study

An analysis of these relationships done by Howell V. Daly and the author has shown, however, that the phylogeny as presented by Winston and Michener (1977) and Kimsey (1984b) and their conclusions as to a dual origin of highly eusocial behavior in the Apidae may not be correct. New evidence indicates that the Meliponini and Apini are sister taxa and that the Euglossini form the sister to the rest of the Apidae as was formerly believed (Figure 3.2d).

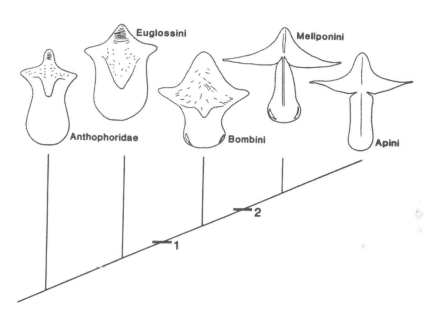

Figure 3.3 Ventral view of prosterna, anterior above. (1) and (2) refer to derived character states on the cladogram: (1) = expansion of basisternum (= triangular anterior region of prosternum) seen in Bombini, Meliponini and Apini, but not Euglossini and outgroup. (2) = constriction formed between the basisternum and furcasternum (= elongated posterior region of prosternum) and the obscuring of the prosternum apophyseal pit (indicated by v-shaped line in middle of illustrations of Anthophoridae, Euglossini and Bombini) seen in the Meliponini and Apini, but not Bombini, Euglossini or outgroup. Cladogram is the same as illustrated in Figure 3.2d.

The support for the sister group relationship of the Meliponini and Apini involves eight new characters of the hypopharyngeal suspensorium, prosternum, maxillus, scutum, and a feature of the cuticle, as well as a few previously known characters such as the lack of rear tibial spurs in these two tribes (Table 3.1A). These characters will be fully described and illustrated in a later paper (Prentice and Daly, in prep.). Of these new characters, the features of the prosternum are particularly strong as synapomorphies and are illustrated here. Both the Meliponini and Apini uniquely possess a characteristic constriction between the triangular anterior region and elongated posterior region of the prosternum (called the basisternum and furcasternum respectively by Snodgrass, 1956) and lack identifiable prosternal apophyseal pits (Figure 3.3). These features of the prosternum provide excellent evidence for the sister group relationship of the Meliponini and Apini as the prosternum is extremely conservative in shape throughout the Apoidea (resembling the condition illustrated for the Anthophoridae) and it is therefore unlikely that the similar conditions of the Meliponini and Apini evolved independently.   No other bee taxa (indeed no other Hymenoptera) have such prosterna.   The other derived features characteristic of the Meliponini and Apini listed in Table 3.1A are unknown or very uncommon in other bees, including the outgroup, and therefore provide strong support for the sister group relationship of the Meliponini and Apini.   Characters that are less variable in the outgroup are more positively polarized as to their ancestral and derived character states within the family and are therefore more useful as indicators of their phylogenetic relationships.

The characters used by Winston and Michener (1977) and Kimsey (1984b) to separate these two tribes and place the Meliponini as the sister to the rest of the Apidae are listed in Table 3.1B.   Of these characters, some are weak as synapomorphies because they are variable in the outgroup.   Some others are erroneous.   For example, Winston and Michener (1977) said that the development of the basistipital process supported the separation of the Meliponini from the other apid tribes.   The absence of the process was supposedly a synapomorphy of the Euglossini, Bombini and Apini as it was reported to be present in the Meliponini and outgroup.   However, as Kimsey (1984b) correctly observed, the basistipital process is lacking only in the Apini.   Therefore the lack of this process is only a derived character of that tribe alone and is not indicative of the tribal relationships within the family.   Likewise, the form of the subgaleal sclerite (stipital sclerite of Winston, 1979) in the Meliponini, which was reported to be similar to the condition in the outgroup and therefore likely ancestral for the Apidae, does not support Winston and Michener's claim. The stipital sclerite is quite variable in form and development in the outgroup and the condition found in the Euglossini is actually quite similar to conditions found in certain members of the outgroup.   Indeed, the stipital sclerite of the Meliponini is most similar in overall shape to that of the Apini, not the outgroup.   The use of the small stigma as a synapomorphy of the Euglossini, Bombini and Apini is also not a good character uniting them.   Not only is the size of the stigma variable in the outgroup and other bees, making the determination of the ancestral apid condition weak at best, the stigma of *Melipona* (now considered to be the basal clade of the Meliponini; C. D. Michener, pers. com.) is small and shaped very much like that of *Apis*.   Consequently, a small stigma could be plesiomorphic for the Meliponini and thus ancestrally characterize all

Table 3.2 Morphological characters uniting the Bombini with the Meliponini and Apini *versus* characters uniting the Bombini with the Euglossini.

Characters uniting Bombini with:

A. Meliponini and Apini	B. Euglossini
1. Postmentum separated	1. Jugal lobe absent or vestigial
2. Basisternum expanded	2. Arolium absent or vestigial
3. Internal metapleural ridge bent	3. Well developed mandibular grooves
4. Smoothly curving ridge of retractor muscle of second phragma	4. Large, robust form
	5. Papillate distal wing membrane
	6. Postgenal lobes free
	7. "Large" prosternum

the apid tribes. Therefore, the stigma does not provide evidence for the sister-group relationship of the Meliponini to the rest of the Apidae.

The strongest support for the sister group relationship of the Meliponini to the other apids is the presence of the auricle in the Euglossini, Bombini and Apini (Winston and Michener, 1977). The auricle is an expansion of the base of the basitarsus of the hind leg which is used in the packing of pollen onto the corbicula of the rear tibia. It is present in all apids except the Meliponini (Michener et al., 1978). The Meliponini pack pollen onto the corbicula, instead, with the penicillum, which is a brush of recurved hairs on the outer anterior corner of the hind tibia (Figure 3.4). The lack of the auricle in the Meliponini has been assumed to be ancestral for the Apidae because the auricle is lacking in the outgroup and other bees. If the Meliponini and Apini are sister taxa, however, the auricle must have been replaced by the penicillum in the ancestors of the Meliponini. This transition is supported by the fact that in the Bombini and Apini there is a brush of unmodified hairs where the penicillum would originate in the Meliponini (Figure 3.4). Likely, the penicillum evolved as a modification of such hairs and if so, probably replaced the auricle as the means of pollen packing in the ancestors of the Meliponini. If this reasoning is correct, the auricle represents a derived character uniting the Apidae. Its loss is only a derived character of the Meliponini, and therefore does not support the separation of this tribe as the sister to the rest of the Apidae.

If the Meliponini and Apini form a holophyletic group, this leaves only two likely cladograms for the relationships within the family, illustrated in Figure 3.2a, d. The difference between these two alternatives is whether the Euglossini and Bombini form a holophyletic group separate from the Meliponini + Apini (Figure 3.2a) or whether the Bombini are more closely related to the Meliponini + Apini than to the Euglossini (Figure 3.2d). The best support for the second alternative is to be found on the postmentum. Plant and Paulus (1987) showed that the Euglossini were the only apid tribe to retain the assumed ancestral condition for the Apidae of a united postmentum as found in the outgroup. The Bombini, Meliponini and Apini all have separated postmenta (Figure 3.5). Because of this, Plant and Paulus (1987) suggested that the Euglossini, not the Meliponini, formed the sister to the rest of the Apidae (Figure 3.2c). Howell V. Daly and the author have found other characters in support of Plant and Paulus's view that the Euglossini are the sister to the other apids (Table 3.2A) which we will fully describe in a later paper. Of these characters, the form of the prosternum is illustrated in Figure 3.3. The Bombini share with the Meliponini and Apini a very well developed basisternum which in the Bombini, at least, serves to separate the propleura, forming a depression for the base of the proboscis. This is a derived character uniting these three tribes because the Euglossini retain the ancestral apid condition of a small basisternum as found in the outgroup and other bees. Like the form of the postmentum, this is strong evidence for the holophyly of these three tribes because no other bees have similarly modified basisterna. Other characters suggest that the Euglossini are less related to the other apids than indicated by the phylogeny of Figure 3.2a. As pointed out by Sakagami and Michener (1987), the Euglossini are the only apids to retain the apparently primitive male metasomal sterna as found in *Manuelia* of the outgroup. They are also dissimilar to the other apids in their glossal structure (Michener and Brooks, 1984) and have the most primitive apid larvae (Zucchi et al., 1969). The auricle of the Euglossini is also quite different from the auricle of other apids (in fact, Wille (1979b) did not consider the auricle to be present in the Euglossini) and might be ancestral for the Apidae, although a positive determination as to the primitive condition in this character cannot be made since no other bees have similar structures for character polarization. Euglossines also do not share certain behavioral characters with the other apids such a food storage outside of brood cells (Sakagami and Michener, 1987).

Those characters supporting the alternative hypothesis (i. e., a close relationship between the Euglossini and Bombini as shown in Figure 3.2a) described by Winston and Michener (1977) and Kimsey (1984b) are listed

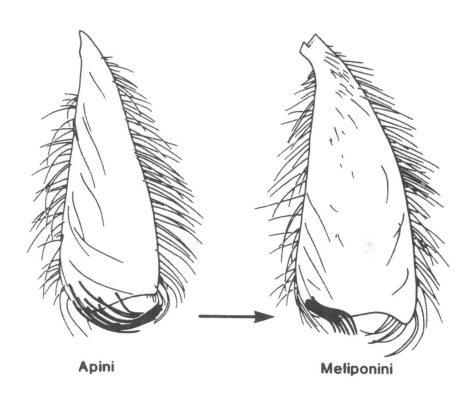

Figure 3.4 Lateral view of left tibia of the Apini (*Apis*) and Meliponini (*Melipona*) showing the hypothesized evolution of the penicillum from apine-like lower corbicular hairs (penicillum = recurved hairs at the bottom of the illustration of the meliponine tibia). In the Apini and Bombini these lower corbicular hairs are undifferentiated from the surrounding simple corbicular hairs and are likely ancestral to the highly modified hairs of the meliponine penicillum. This demonstrates that the evolution of the penicillum from the apine or bombine condition is plausible. The corbicula, or pollen basket, is formed by the smooth cuticle and surrounding hairs of the tibia.

in Table 3.2B. Like the characters which were reported to be primitive in the Meliponini, many of these characters are weak or inaccurate. For example, the incomplete hypostomal bridge and enlarged prosternum do not represent synapomorphies of these two tribes. Instead, the conditions evident in the Euglossini in these characters are similar to conditions evident in the outgroup and are therefore likely ancestral for the Apidae. Other characters such as size, papillae of distal wing membrane and mandibular grooves, are weak not only because they are known to be variable in other bees (for example, the outgroup ranges from very small bees such as *Ceratina* to quite large ones such as *Xylocopa*) but also because they may be correlated with each other and may just as likely be ancestral for the Apidae. Indeed, if the Euglossini do form the sister to the rest of the Apidae, most of the characters that have been used to unite the Euglossini and Bombini probably represent ancestral character states for the Apidae. Apparently, the only strong evidence for their holophyly comes from the loss of the arolium and jugal lobe in these two taxa. It should be noted, however, that even with regard to these characters, they are losses and are therefore more subject to convergence. The arolium, in particular, has been lost independently in many bee taxa (including the outgroup) and may have been lost independently in the Euglossini and Bombini. It should be added that the conditions of these features are not quite identical in these two tribes. The arolium is absent in the Euglossini, vestigial but present in the Bombini while the jugal lobe is absent in the Bombini, vestigial but present in the Euglossini. This supports the idea that these characters have been reduced independently in these two tribes and therefore do not represent synapomorphies of them.

Therefore, on morphological grounds the Euglossini probably form the sister to the rest of the Apidae while the Bombini form the sister to the Meliponini + Apini (Figure 3.2d).

Sociality in the Apidae

Of all the possible phylogenetic schemes relating the tribes of Apidae, the phylogeny of Figure 3.2d fits best with the distribution of sociality in the family. In this scheme, only a single origin of primitive eusociality and then high eusociality needs to be assumed (Figure 3.6a). This contrasts with the phylogeny of Figure 3.2b where many more independent origins of eusociality are required. It is assumed that the evolution of high eusociality requires first the evolution of primitive eusociality. Therefore, an evolution of high eusociality from a solitary ancestor requires two character state changes on a cladogram instead of a single character state change. Thus, if the common ancestor of the Apidae was social only to the degree of the Euglossini, as Winston and Michener (1977) suggested, then three independent origins of primitive eusociality and two independent origins of high eusociality would need to be assumed (Figure 3.6b) using the phylogeny of Figure 3.2b. Clearly this is a less parsimonious and therefore less probable scheme than that suggested by Figure 3.6a (five character state changes instead of two on the cladogram). A more

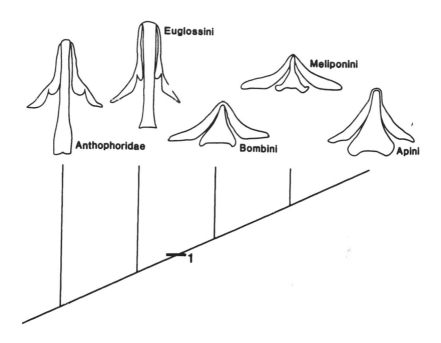

Figure 3.5 Posterior view of postmenta. (1) refers to the derived character state on the cladogram of a separation between the lorum and mentum seen in the Bombini, Meliponini and Apini but not the Euglossini or outgroup (Plant and Paulus, 1987). Lorum = inverted v-shaped sclerite, mentum = elongated or triangular shaped sclerite directly below the lorum. In the Anthophoridae and Euglossini these sclerites are continuous above. Cladogram is the same as illustrated in Figure 3.2d.

parsimonious arrangement for the phylogeny of Figure 3.2b would involve a primitively eusocial common ancestor to the Apidae, two independent origins of high eusociality in the Meliponini and Apini and a loss of primitive eusociality in the line leading to the Euglossini (Figure 3.6c). This is still less parsimonious than the first scheme (four character state changes versus two on the cladogram), and it should be noted that both a dual origin of high eusociality in the Mcliponini and Apini and a loss of primitive eusociality in the Euglossini are unlikely events. Thus, the distribution of sociality in the family strongly argues for the phylogenetic scheme of Figure 3.2d.

Therefore, high eusociality probably arose only once in the family contrary to the conclusions of Winston and Michener (1977). While differences do exist between the Meliponini and Apini in their social systems, they may be explained as due to evolutionary divergence from a highly eusocial common ancestor (some differences are listed in Table 3.3). For example, with respect to nest architecture, the horizontal, single sided combs of the Meliponini likely represent the ancestral state for the vertical, double sided apine combs. This is because brood cells of the Euglossini and Bombini are usually vertically oriented and are more similar to those of the Meliponini than those of the Apini. Horizontal, single-sided combs with vertical cells are a more natural progression from these largely independent vertically oriented brood cells (Figure 3.7a to b). The fact that horizontal combs are sometimes found in the Euglossini (Zucchi et al., 1969) (assuredly evolved independently of the Meliponini) supports this contention. That Apine comb may have evolved from the horizontal combs of the Meliponini is supported by the independent evolution of vertical, double-sided combs in a single species of Meliponini, *Dactylurina staudingeri* (Wille and Michener, 1973), demonstrating that such a transition can occur (Figure 3.7b to c). The mass provisioning found in the Meliponini is also likely ancestral to the progressive provisioning of the Apini since almost all bees mass provision their cells. This indicates that this is the more primitive characteristic. It should be added that the provisions of the Meliponini and Apini are unique among bees in being highly supplemented with adult pharyngeal gland secretions (Michener, 1974) which adds further support to the contention that these two tribes are sister taxa. Their difference in colony founding by new queens in the Meliponini and old queens in the Apini may likewise be logically accounted for as due to a transition from one to the other character states in the ancestors of one or the other tribe. That a transition may have occured in colony founding is demonstrated by the fact that a species of Meliponini (*Tetragona laeviceps*) has been reported to abscond with the old queen (Inoue et al., 1984), and this is in some ways similar to the founding of colonies by old queens in *Apis* and may be a remnant of such behavior. The other differences between the social systems of the Meliponini and Apini listed in Table 3.3 may be similarly explained.

Ideas on the Evolution of the Apidae and Conclusion

If the phylogeny of Figure 3.2d represents the relationships within the family, it is likely the progression from Anthophoridae to Euglossini to Bombini and finally to Meliponini and Apini parallels the morphological and behavioral evolution of the family. In many respects, the Euglossini represent a "link" between the Anthophoridae and other Apidae. As evidenced by their former placement with the anthophorids, they are

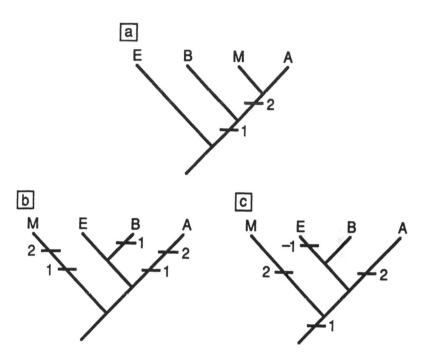

Figure 3.6 Character state changes in social behavior on the cladograms proposed by Winston and Michener (1977) (b and c) and the present paper (a). E = Euglossini, B = Bombini, M = Meliponini, A = Apini, 1 = evolution of primitive eusociality, 2 = evolution of high eusociality, -1 = loss of primitive eusociality. The total character state changes shown in diagram (a) = 2, diagram (b) = 5, and diagram (c) = 4.

similar to them in a number of ways. They share certain morphological details with that family (Plant and Paulus, 1987; Prentice and Daly, in prep.) and in their overall appearance and characteristic swift flight are reminiscent of some larger anthophorids. They are also basically solitary like most anthophorids. They share with the other apids, however, the derived characters of the family which demonstrate their close relationship with them. Of the apid tribes, they are most similar to the common ancestor of the family while not identical to this common ancestor since they show a number of their own uniquely derived characters, as mentioned above. The Bombini are intermediate between the largely solitary Euglossini and the highly eusocial Meliponini and Apini. In overall appearance, such as their large size and general hairiness, the Bombini more resemble some Euglossini than the small and less haired Meliponini and Apini. In their sociality, however, they are more similar to the Meliponini and Apini, having morphologically differentiated queen

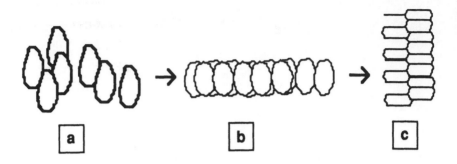

Figure 3.7 Evolution of the comb of the Meliponini and Apini: (a) = clumped, vertically oriented brood cells seen in the Euglossini and Bombini and some Meliponini, (b) = horizontal, single sided combs seen in the Meliponini, (c) = vertical, double sided combs seen in the Apini and one species of Meliponini (*Dactylurina staudingeri*).

and worker castes. Probably in an ancestor very similar to the present day Bombini, the line leading to the Meliponini and Apini evolved high eusociality. When high eusociality evolved, a smaller, less haired bee was selected for as large, highly eusocial colonies afforded greater protection from the environment. Queen and worker castes became more specialized in their respective tasks with queens losing the ability to forage and workers losing the ability to effectively reproduce as in modern day Meliponini and Apini. In some respects the Bombini are structurally intermediate between the Meliponini and Apini and this supports the hypothesis that these two tribes were ultimately derived from a bombine-like ancestor. For example, the Bombini have both dorsal and ventral wax glands on their metasoma while the Meliponini have only dorsal glands and the Apini only ventral ones. It is likely that this difference resulted from the loss of the dorsal glands in the Apini and the ventral ones in the Meliponini from a common ancestor with both dorsal and ventral glands, as in the Bombini (Cruz Landim, 1963). Morphologically, the immediate common ancestor of the Meliponini and Apini may have been somewhat intermediate between these two tribes. This is because both tribes have a number of highly derived characters that represent morphological evolution away from their common ancestor. Such features as the penicillum and reduced wing venation of the Meliponini and the highly modified male genitalia of the Apini are examples. That their common ancestor was intermediate between them is supported by the fact that the fossil apine, *Electrapis*, (considered to be ancestral to *Apis* by

Table 3.3 Some differences in the social systems between the Meliponini and Apini.

	Meliponini	**Apini**
Comb orientation:	Horizontal, single-sided combs (one species with vertical double-sided combs)	Vertical, double-sided combs
Honey and pollen storage:	In specialized storage pots	In comb
Worker communication for food:	By smell, and other ways	By smell, round and waggle dances and other ways
Reuse of brood cells:	Not reused	Reused
Founding of new colonies:	By new queens in swarms	By old queens in swarms
Feeding of immatures:	Mass provisioned	Progressively provisioned

some [Zeuner and Manning, 1976]) is quite meliponine-like in overall appearance and has even been classed within the Meliponini (Kerr and da Cunha, 1976). In their sociality, however, the Meliponini are probably much closer to the common ancestor of both tribes than are the Apini, and likely display many of the characteristics in social behavior of the ancestors of *Apis*.

Future studies concerning the evolution within *Apis* should be benefited by considering these ideas on the relationship of *Apis* to the other Apidae. Since the Meliponini now likely form the sister to *Apis*, this tribe should form the main "outgroup" in evaluating ideas on the evolution of structures and behaviors within the genus. Quite possibly, the origin and evolution of such clearly derived traits as the dance language and other behaviors of *Apis* may be ultimately found by study of representatives of the Meliponini. The phylogeny of Figure 3.2d provides a framework for evaluating such ideas on the evolution of *Apis* and other members of the Apidae.

Acknowledgments

I thank Howell V. Daly who initiated this study of the relationships within the Apidae and for valuable discussions concerning a closely related manuscript. I would also like to thank Charles D. Michener for discussions concerning the related manuscript. I also thank Cristina Jordan for inking the illustrations of the prosterna, postmenta and tibiae (Figures 3.3, 3.4, 3.5).

Literature Cited

Ashmead, W. H. 1899. Classification of the bees or the superfamily Apoidea. *Trans. Am. Entomol. Soc.* 26:49-100.

Cruz Landim, C. da. 1963. Evolution of the wax and scent glands of the Apinae. *J. NY Entomol. Soc.* 71:2-13.

Inoue, T., S. F. Sakagami, S. Salmah, and Nismah Nukmal. 1984. Discovery of successful absconding in the stingless bee *Trigona (Tetragonula) laeviceps. J. Apic. Res.* 23:136-142.

Kerr, W. E., and R. A. da Cunha. 1976. Taxonomic position of two fossil social bees (Apidae). *Rev. Biol. Trop.* 24:35-43.

Kerr, W. E. and V. Maule. 1964. Geographic distribution of stingless bees and its implications (Hymenoptera: Apidae). *J. NY Entomol. Soc.* 72:2-17.

Kimsey, L. S. 1984a. The behavioral and structural aspects of grooming and related activities in euglossine bees. *J. Zool. (Lond).* 204:541-550.

------. 1984b. A re-evaluation of the phylogenetic relationships in the Apidae (Hymenoptera). *Syst. Entomol.* 9:435-441.

------. 1987. Generic relationships within the Euglossini (Hymenoptera, Apidae). *Syst. Entomol.* 12:63-72.

Latreille, P. A. 1802. *Histoire naturelle, générale et particulière des crustaces et des insectes. 3.* Paris. xii + 467 pp.

Lepeletiere de S. Fargeau, A. 1836. *Histoire naturelle des insectes, hyménoptères. 1.* Roret, Paris. Pp. 1-547.

Maa, T. 1953. An inquiry into the systematics of the tribus Apidini or honeybees (Hym.). *Treubia* 21:525-640.

Michener, C. D. 1944. Comparative external morphology, phylogeny, and a classification of the bees (Hymenoptera) *Bull. Am. Mus. Nat. Hist.* 82:151-326.

------. 1958. The evolution of social behavior in bees. *Proc. Xth Internat. Congr. Entomol.* [Montreal] 2:441-448.

------. 1964. Evolution of the nests of bees. *Am. Zoologist* 4:227-239.

------. 1972. Direct food transferring behavior in bees. *J. Kans. Entomol. Soc.* 45:373-376.

------. 1974. *The Social Behavior of the Bees.* Harvard University Press: Cambridge, Mass.

Michener, C. D., M. L. Winston, and R. Jander. 1978. Pollen manipulation and related activities and structures in bees of the family Apidae. *Univ. Kans. Sci. Bull.* 51:575-601.

Michener, C. D. and R. W. Brooks. 1984. Comparative study of the glossae of bees (Apoidea). *Contrib. Am. Entomol. Inst.* 22:i-iii+1-73.

Plant, J. D. and H. F. Paulus. 1987. Comparative morphology of the postmentum of bees (Hymenoptera: Apoidea) with special remarks on the evolution of the lorum. *Z. zool. Syst.. Evolutionsforsch.* 25:81-103.

Roubik, D. W. 1989. *Ecology and Natural History of Tropical Bees.* Cambridge University Press: Cambridge.

Ruttner, F. 1988. *Biogeography and Taxonomy of Honeybees.* Springer-Verlag: New York.

Sakagami, S. F. 1971. Ethosoziologischer Vergleich zwischen Honigbienen und stachellosen Bienen. *Z. Tierpsychol.* 28:337-350.

Sakagami, S. F., and C. D. Michener. 1987. Tribes of Xylocopinae and origin of the Apidae (Hymenoptera: Apoidea). *Ann. Entomol. Soc. Am.* 80:439-450.

Snodgrass, R. E. 1956. *Anatomy of the Honey Bee.* Cornell University Press: Ithaca NY

Williams, P. H. 1985. A preliminary cladistic investigaion of relationships among the bumble bees (Hymenoptera, Apidae). *Syst. Entomol.* 10:239-255.

Wille, A. and C. D. Michener. 1973. The nest architecture of stingless bees with special reference to those of Costa Rica, (Hymenoptera: Apidae). *Rev. Biol. Trop.* 21 (supl. 1.)

Wille, A. 1979a. Phylogeny and relationships among the genera and subgenera of the stingless bees (Meliponinae) of the world. *Rev. Biol. Trop.* 27:241-277.

------. 1979b. A comparative study of the pollen press and nearby structures in the bees of the family Apidae. *Rev. Biol. Trop.* 27:217-221.

Winston, M. L. 1979. The proboscis of the long-tongued bees: A Comparative study. *Univ. Kans. Sci. Bull.* 51(22):631-667.

------. 1987. *The Biology of the Honey Bee.* Harvard University Press: Cambridge MA.

Winston, M. L. and C. D. Michener. 1977. Dual origin of highly social behavior among bees. *Proc. Natl. Acad. Sci. USA,* 74: 1135-1137.

Zeuner, F. E. and F. J. Manning. 1976. A monograph on fossil bees (Hymenoptera: Apoidea). *Bull. Brit. Mus. (Nat. Hist.) Geol.* 27:149-268.

Zucchi, R., S. F. Sakagami, and J. M. F. Camargo. 1969. Biological observations on the neotropical parasocial bees, *Eulaema nigrita,* with review on the biology of Euglossinae. *Journal of the Faculty of Science Hokkaido University, Zoology* 17:271-380.

4

A New Tribal Phylogeny of the Apidae Inferred from Mitochondrial DNA Sequences

Sydney A. Cameron

Introduction

The genus *Apis* (honey bees) has evolved some of the most complex social behavior of any of the Apidae (reviews in Seeley, 1985; Winston, 1987). Its symbolic dance language (Gould et al., 1985, Dyer, 1987) and polyandrous mating system (Page, 1980) are unique among bees. Two other behavioral features found thus far only in *Apis mellifera*, though little investigated in other species (but see Cameron and Robinson, 1990), are a hormonally driven age polyethism (Robinson, 1985, 1987) and a reproductive physiology that appears to be independent of juvenile hormone (Robinson et al., in review).

A second apid lineage, the Meliponini (stingless bees), displays a highly eusocial level of colony organization comparable to that of *Apis* species. However many characteristics of stingless bee social biology differ in detail from those of honey bees, raising a question as to whether these two forms of highly eusocial behavior in the Apidae are truly homologous. Are these two forms of sociality variations upon a highly eusocial theme that developed early in the evolution of the Apidae, or are they independently derived phenomena convergent in their level of social organization? To put highly eusocial behavior into an evolutionary context, it is important to have a reliable phylogeny, not only for the genus *Apis* but for the whole of the Apidae.

71

In considering apid social behavior in an evolutionary context, we are fortunate that Apidae contains taxa exhibiting all levels of social organization, from solitary (most Euglossini), to primitively eusocial (probably all Bombini), to highly eusocial (Apini and Meliponini). In this context, there is great potential for phylogenetic analysis of comparative behavioral traits to elucidate the evolution of highly eusocial behavior in the Apidae.

The Phylogenetic Approach to Bee Behavior

A resurgence of the phylogenetic approach among honey bee behaviorists and ecologists is evident in current investigations of the dance language (Dyer and Seeley, 1989), mating behavior (Koeniger and Koeniger, 1990), orientation behavior (Jander, 1976), and nesting and defense strategies (Seeley, 1985). In fact, analysis of behavioral character evolution in a phylogenetic context, *sensu* Remane (1952), is reemerging throughout the discipline of comparative behavior (e.g., Clutton-Brock and Harvey, 1984; Greene, 1986; Huey and Bennett, 1987; Coddington, 1988; McLennan et al., 1988; Sillén-Tullberg, 1988; Carpenter, 1989; Baum and Larson, 1991; Brooks and McLennan, 1991). This new emphasis on the comparative method is not entirely stochastic; it appears to be associated with rigorous new methods for inferring phylogenies (Felsenstein, 1981; Hendy and Penny, 1982; Farris, 1983, 1988; Felsenstein, 1988a, b; Swofford, 1990) and estimating their reliability (Templeton, 1983; Felsenstein, 1985a, b; Templeton, 1987; Sanderson, 1989). Hand in hand with the new comparative trend is the realization that behavioral traits alone cannot resolve questions of behavioral character evolution because evolutionary hypotheses of behavior require independently estimated phylogenies as a foundation for their testing (e.g., Ridley, 1983; Larson, 1984; Felsenstein, 1985c; Greene, 1986; Huey, 1987; Pagel and Harvey, 1988; Donoghue, 1989; Maddison, 1990; Mickevich and Weller, 1990; Baum and Larson, 1991).

To establish the ancestral state of a trait that evolves within a group, one must have information on the expression of that trait outside the group, among its closest relatives (Watrous and Wheeler, 1981; Donoghue and Cantino, 1984; Maddison et al., 1984; Swofford and Maddison, 1987). For example, an investigation of the dance language within *Apis* requires knowledge of the homologous traits among the closest relatives of *Apis* in the family Apidae.

In this chapter I review previously proposed phylogenies of the tribes of Apidae, with particular emphasis on their implications for the evolution of highly eusocial behavior. I also present the results of my own analysis, based on an independent data set obtained from mitochondrial DNA (mtDNA) sequences. The results of the mtDNA analysis lead to a new hypothesis of apid relationships that is strongly discordant with hypotheses based on morphology. Finally, I discuss the implications of this new hypothesis for the evolution of social behavior in the Apidae.

Morphology-Based Phylogenies for the Apidae

Phylogenetic relationships among the tribes of Apidae traditionally have been inferred by comparative study of morphological characters. An overview of the morphological characters of each tribe and their implications for phylogenetic relationships has been prepared by Prentice (Chapter 3, this volume). For another treatment of tribal morphology, see Michener (1990a).

There are four major hypotheses of relationships among the apid tribes, all based on internal or external morphological traits. Michener (1944) first proposed that highly eusocial behavior arose once in the ancestor of Meliponini and Apini, ultimately from a primitively eusocial common ancestor shared with Bombini (Figure 4.1A). This scenario recently has been corroborated by Prentice (Chapter 3, this volume; Prentice and Daly, in prep) who added several new external and internal skeleto-muscular characters.

The second hypothesis, advanced by Michener in 1974 (Michener, 1974), retains the proposed monophyly of the highly eusocial clade (Meliponini+Apini) but suggests that there is no primitively eusocial common ancestor shared with Bombini. Instead, Bombini is the sister taxon to Euglossini (Figure 4.1B). A recent reanalysis by Michener (1990a) using cladistic methodology was unable to distinguish between these first two hypotheses.

The third hypothesis, advanced by Winston and Michener (1977) and later supported by Kimsey (1984) following reinterpretation of some of their morphological evidence and addition of new external and internal characters, suggests two independent parallel origins for the highly eusocial taxa (Figure 4.1C). Meliponini is considered more primitive in this scheme, sharing several features with the outgroup to the Apidae, the subfamily Xylocopinae in the family Anthophoridae. Furthermore, this scheme retains the sister-group relationship between Euglossini and Bombini.

The fourth hypothesis, advanced by Plant and Paulus (1987), proposes that all the social taxa form a monophyletic unit (Figure 4.1D). Their evidence is based on the derived state of the postmentum shared by Meliponini, Apini, and Bombini. Eusocial behavior is again considered to have arisen only once for the clade. In this respect, their hypothesis is similar to those of Michener (1944) and Prentice and Daly (in prep.). Plant and Paulus depart from these two hypotheses however, by splitting the highly eusocial clade, making Bombini+Apini a monophyletic group. In this scheme, highly eusocial behavior arose twice within the eusocial clade.

A review of these phylogenetic investigations indicate that analyses considering only morphological characters have not come to any consensus regarding relationships in Apidae. Although in some cases apid taxonomists agree on the same topology (e.g., Kimsey, 1984 and Winston and Michener, 1977; or Michener, 1990a and Prentice and Daly, in prep), they disagree with one another's interpretation and choice of characters. Characters are criticized as weak, erroneous, inaccurate, or lacking homology. For the behaviorist trying to understand the evolution of the dance language, for example, this lack of consensus leads to confusion. What criteria are we to use for selecting the best tree among these different hypotheses? Thus far, discrete, phylogenetically informative (synapomorphic) morphological characters for the apids are simply lacking in sufficient quantity to strongly corroborate the tribal relationships.

Rationale for a Molecular Phylogeny of the Apidae

Studies of highly eusocial taxa considering data obtained from gel electrophoresis of enzymes have been restricted to questions of relationship within *Apis* (reviewed in Sheppard and Berlocher, 1989), and have had limited success in resolving relationships. These studies suffer from the same problem that often plagues morphological analyses, a lack of sufficient characters to corroborate relationships.

In recent years, newer methods to assess phylogenetic relationships have developed, revealing new lines of evidence. In particular, classes of DNA have provided alternative characters for analysis (e.g., Avise et al., 1987; Moritz et al., 1987; Harrison, 1989; Goodman, 1989; Hillis and Moritz, 1990). Patterns of restriction fragment length and site polymorphisms of mtDNA have been used in several systematic studies of social Hymenop-

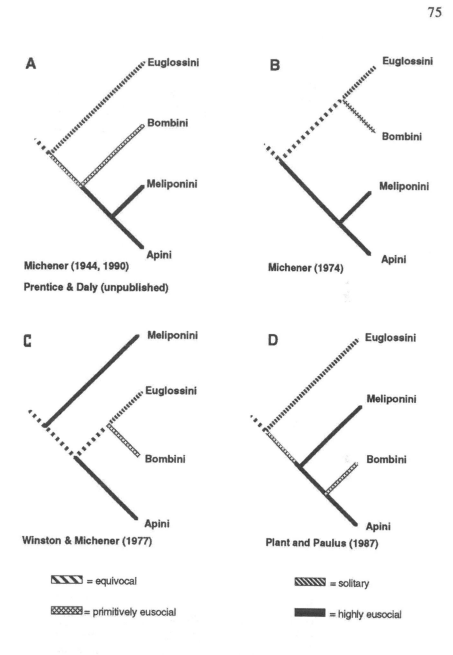

Figure 4.1 Four alternative hypotheses of relationship based on morphological analyses of the four tribes of Apidae.

tera (e.g., Smith, 1991; Smith and Brown, 1990; Smith et al., 1991; Schmitz and Moritz, 1990). The advantages of molecular, especially DNA sequence data, for phylogenetic investigations are as follows.

1. Molecular data provide a new data set, independent from previous analyses and also independent from any morphological or behavioral hypothesis being tested. The importance of having a data set independent from the characters whose evolutionary transformation is being investigated (i.e., target characters) has been questioned recently. De Queiroz (1989) has made the argument that inclusion of target characters for phylogeny reconstruction does not necessarily lead to circular reasoning. However, having an independently derived molecular phylogeny will avoid any analytical problems arising from character covariation or lack of character independence.
2. Nucleotide sequence characters are more likely to be independent of one another than are morphological characters (Baum and Larson, 1991). Character independence is an important assumption in parsimony analysis (Farris, 1983).
3. Aligned sequences of nucleotides provide directly homologous characters for analysis. This is particularly true for mtDNA (discussed below).
4. Sequence characters provide a more or less unlimited supply of additional characters applicable for analyses at all taxonomic levels, from the population to the Kingdom. In other words, some regions of DNA are evolutionarily conserved and others show rapid divergence. Conserved regions are analyzed for higher level taxonomic analyses, whereas regions showing rapid divergence are analyzed for population-level analyses. Furthermore, DNA can be analyzed from the nuclear genome or the mitochondrial genome, or both. For an excellent review of some of these issues see Hillis and Moritz (1990).

The disadvantages of molecular data for phylogenetic analysis are that (1) few genes ahve been studied, hence we lack a large comparative data base for testing hypotheses, and (2) with highly diverged taxa, homoplasy (convergent nucleotide substitutions) among characters increases. Time and effort should take care of the first problem. The second has been addressed by applying various weighting schemes to characters (e.g., Templeton, 1983, 1987; Felsenstein, 1988b).

Although the usefulness of nucleotide sequences for phylogenetic analysis has become widely recognized, the time and expense involved in obtaining the data has prevented widespread use of the technology for systematics. These problems have, in principle, been solved by the revolutionary new development of automated technology for gene amplification based on the polymerase chain reaction (PCR). PCR is a thermocyclic reaction that generates multiple copies of a fragment of DNA relatively quickly and cheaply, eliminating the lengthy procedures of viral and bacterial cloning (Saiki et al., 1985; Mullis et al., 1986; Mullis and Faloona, 1987; Innis et al., 1990). Although PCR is still in its infancy as a tool for systematics, this method now makes it feasible to obtain large quantities of homologous DNA for direct sequencing from individual insects (Wheeler, 1989; Simon et al., 1990; Simon et al., 1991; Cameron, in prep.).

In summary, the molecular approach is potentially powerful for estimating phylogenies and testing competing phylogenetic hypotheses. I have used this approach to examine apid tribal relationships.

Mitochondrial DNA Analysis of Apid Tribal Relationships

MtDNA has become the molecule of choice for sequencing in an increasing number of phylogenetic studies (e.g., Brown et al., 1982; Hixon and Brown, 1986; reviewed in Hillis and Moritz, 1990). Unlike nuclear DNA, mtDNA is maternally inherited (Lansman et al., 1983; Gyllensten et al., 1985) and undergoes little or no recombination (Brown, 1985) or gene duplication (reviewed in Moritz et al., 1987). Thus mtDNA sequences are more easily homologized among different species. Moreover, because this molecule includes both conserved regions (Hixon and Brown, 1986) and highly divergent regions (Fauron and Wolsten-holme, 1980; Aquadro and Greenberg, 1983; Brown, 1985; reviewed in Moritz et al., 1987, pp. 270-271), it is useful for systematic studies at several taxonomic levels.

MtDNA was used in this analysis. Exemplars were selected from each of the four apid tribes: (1) Euglossini (*Eulaema polychroma, Eufriesia caerulescens*); (2) Bombini (*Bombus pennsylvanicus, Bombus impatiens, Psithyrus variabilis*); (3) Meliponini (*Melipona compressipes, Trigona pallens, Trigona hypogea, Scaptotrigona luteipennis*); and (4) Apini (*Apis mellifera, A. cerana, A. dorsata, A. florea*). Based on strong similarities in morphology, the nearest relatives of the Apidae are thought to be from the subfamily Xylocopinae (Sakagami and Michener, 1987). This subfamily, although mostly solitary, contains taxa which exhibit tendencies

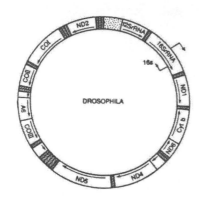

Figure 4.2 Mitochondrial DNA map of *Drosophila* indicating gene order. Inside arrows indicate direction of transcription; the two outside arrows correspond to the position and direction of extension of the oligonucleotide primers used in PCR and sequencing reactions for this study. Redrawn from Clary and Wolstenholme (1985).

toward sociality (Michener, 1990b). The exemplar *Xylocopa virginica* was chosen as a member of the outgroup Xylocopinae.

The characters considered in my analysis are based on nucleotide sequences of mtDNA from individual thoraces. To obtain the sequences, total genomic DNA was extracted from thoracic tissue and purified using modifications (Cameron, in prep.) of standard procedures (Maniatis et al., 1982). MtDNA was then amplified via PCR, with a set of primers based specifically on sequences from bumble bee and honey bee large (16s) rRNA genes (Figure 4.2). These primers share sufficient homology with all the apid tribes to amplify a 600 bp fragment near the 3' end of this gene (Figure 4.2). The 16s rRNA gene was chosen because it is known to contain regions of conserved nucleotides (Gerbi, 1985).

Amplified double stranded mtDNA was then sequenced directly using modifications of the methods of Sanger (Sanger et al., 1977) and Dubose (DuBose and Hartl, 1990). The sequences for each taxon were compared and the homologous regions aligned by hand in order to determine whether there were genetic differences between taxa. Each nucleotide position is a potential character for phylogenetic analysis, with the different nucleotides expressed by each taxon at a given site representing different character states. A total of 550 nucleotides were scored for each of the 14 taxa described above. Homologizable nucleotide substitutions constituting informative characters were identified at 129 sites out of the total 550. Of these, 27 were informative in determination of tribal relationships (see Figure 4.3). A more complete description of the methods and nucleotide sequence data will be presented elsewhere (Cameron, in prep.).

Parsimony analysis, using the branch and bound method implemented in PAUP version 3.0L (Swofford, 1990), was applied to the 129

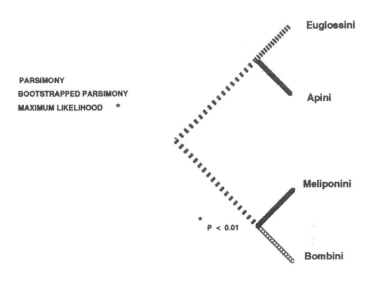

Figure 4.3 A maximum parsimony tree (inferred from comparisons of nucleotide sequences of mtDNA [16s rRNA] from 14 taxa), simplified to show only the tribal topology. Tree length = 353 steps, consistency index = 0.60 (calculations based on 14 taxa). Maximum Likelihood analysis produced the same tree. The * and its associated *p*-value indicates the statistical significance of the Meliponini+Bombini clade as calculated under the Maximum Likelihood model. Of the 13 synapomorphies uniting Apini+Euglossini, seven were homoplasious with members of the sister clade. Four of the 14 synapomorphies uniting Bombini+Meliponini were homoplasious with the sister clade. Overall for the tribes, 26 synapomorphies united the species of Apini (15 of these were homoplasious, i.e, occurred elsewhere on the cladogram), 11 united Euglossini (9 homo.), 24 united Meliponini (14 homo.), and 13 united Bombini (12 homo.). Social status of the tribes as in Fig. 4.1.

phylogenetically informative characters. The result of this analysis was four equally parsimonious trees exhibiting the same tribal topology, with one clade composed of the sister groups Apini and Euglossini, and the other clade composed of the sister groups Bombini and Meliponini (Figure 4.3). A second topology (Figure 4.4), one step longer than the most parsimonious trees, shows a Bombini+Meliponini clade, with Apini as the sister group to this clade and Euglossini as the sister group to Apini+Bombini+ Meliponini. These results do not agree with the previous phylogenies of Apidae based on morphology (Figures 4.1A-D). In particular, no morphological analysis has ever indicated that Bombini and Meliponini form a monophyletic group. To estimate the reliability of this clade, I applied the bootstrap method (Felsenstein, 1985b) implemented in

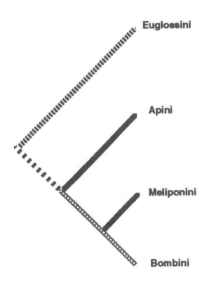

Euglossini

Apini

Meliponini

Bombini

Figure 4.4   Alternative tree, 1 step longer than the most parsimonious tree (Figure 4.3). Social status depicted as in Figure 4.1.

PAUP (Swofford, 1990). Bootstrap randomly resamples characters (with replacement) from the data set and recalculates a tree for each resample. The Bombini+Meliponini clade appeared in 86 out of 100 phylogenies calculated from 100 resamplings of the data. Although not a statistically significant result, it indicates a high degree of corroboration among the characters supporting this clade.

Some forms of nucleotide substitutions in mtDNA occur at higher frequencies than others, exhibiting a greater number of transitions than transversions, or *vice versa* (Brown et al., 1982; Aquadro and Greenberg, 1983; Greenberg et al., 1983; Wolstenholme and Clary, 1985; Satta et al., 1987). To account for any bias in my data, a model of sequence evolution derived empirically from the data was used in a Maximum Likelihood analysis (Felsenstein, 1981). The transition/transversion bias calculated from the data was approximately 3:1. A subset of the taxa (two representatives from each tribe plus the outgroup) was analyzed (implemented in PHYLIP, Felsenstein,1988a).

Maximum Likelihood produced the same basic tree (Figure 4.3). Based on this model, the branch length supporting the Bombini+Meliponini clade was significantly different from random ($p \leq 0.01$), while the branch length supporting the Apini+Euglossini clade was not significantly different from random ($p \geq 0.05$).

Because the results of this molecular analysis differ so strikingly from the previous morphological analyses, it is appropriate to ask whether the topology calculated here (Figure 4.3) is better in explaining my own data than those proposed by previous investigators (Figure 4.1A, C, and D). Using two different sets of exemplars from each tribe, the length for each of the four topologies (Figures 4.3, 4.1A, C, and D) was calculated. The tree based on molecular data was seven steps shorter (173 steps) than that shown in Figure 4.1A (180), nine steps shorter than C (182), and 14 steps shorter than D (187).

To test whether a topology showing a Bombini+Meliponini clade is significantly shorter than the alternatives, I applied Felsenstein's test for molecular phylogenies (Felsenstein, 1985a). This test compares all topologies among three taxa only, using exclusively phylogenetically informative characters for the taxa. Because the distribution of highly eusocial behavior is a primary concern for all of the topologies discussed above, I specified all of the possible three-taxon arrangements for the social taxa (Apini, Meliponini, and Bombini), excluding the solitary Euglossini (Figures 4.5A, B, and C). The topology of Figure 4.5B, indicating a Bombini+Meliponini clade, is

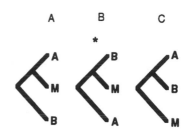

Figure 4.5 Comparison of the three possible topologies for the social tribes (A = Apini, M = Meliponini, and B = Bombini). Application of Felsenstein's test (1985a) to the null hypothesis that all three topologies are equally supported resulted in falsification of H0. Topology A is supported by 10 informative sites, topology B by 19 sites, and topology C by 7 sites. Topology B was significantly better supported (* = $p <$ 0.05). This is consistent with the hypothesis that Bombini and Meliponini are more closely related than either is to Apini.

supported by 19 phylogenetically informative characters, and is nine steps shorter (53 steps) than the next shortest alternative, an Apini+Meliponini clade (62 steps, Figure 4.5A), and 12 steps shorter than Figure 4.5C, an Apini+Bombini clade. The tree in Figure 4.5B represents a significantly better topology than the other two ($p < 0.05$).

Conclusion

The primary aim of this chapter is to report a new tribal phylogeny for the Apidae based on nucleotide sequences of mtDNA. Results of the above analyses (Figure 4.3) indicate that Apini and Euglossini are sister groups and Meliponini and Bombini are sister groups. However, the tree in Figure 4.4 should be kept in mind as an alternative hypothesis, considering the higher level of homoplasy among the synapomorphies for the Apini+Euglossini clade. This topology maintains the Meliponini+Bombini clade, but suggests that Apini is the sister to the Meliponini+Bombini clade. W. S. Sheppard and B. A. McPheron (Chapter 5, this volume) have independently obtained data that support the Meliponini+Bombini clade.

Their study, which represents the only other attempt to analyze apid relationships using genetic data to date, was based on analyses of sequences from the 18s nuclear rRNA for the following taxa: *Bombus terricola, Apis mellifera, Euglossa imperialis, Trigona capitata, Anthophora abrupta,* and *Xylocopa virginica).* The most parsimonious tree based on mtDNA nucleotide sequences has thus been independently corroborated by nucleotide sequences from the nuclear genome.

To resolve the residual uncertainty between the two trees in Figures 4.3 and 4.4, future work should consider other nucleotide sequences from both mitochondrial and nuclear genes. It would be especially useful to incorporate evolutionarily conservative regions to avoid high levels of homoplasy. There is controversy over the issue of character independence for genetic characters (Harrison, 1989; Hillis and Moritz, 1990), just as for morphological characters. Incorporating data from other regions of the genome will ensure a relatively high level of independence for character analysis. Furthermore, additional data will strengthen the estimates of phylogenetic reliability discussed above, thus providing a more powerful basis for testing one hypothesis against its alternatives.

These molecular analyses lead to a fascinating alternative framework from which to interpret the evolution of specific social traits in Apidae. All of my analyses indicate that Meliponini and Bombini are sister groups, but the position of Apini as sister group to Euglossini requires further corroboration. It is evident from these results that if highly eusocial behavior is derived relative to primitively eusocial behavior, it arose twice independently, once in the meliponine lineage and again within the apine lineage. This dual origin of high eusociality is implied in both the favored topology (Figure 4.3) and the next best alternative (Figure 4.4). This would suggest that highly eusocial traits of honey bees and stingless bees, such as components of the dance language, may not be homologous unless they can be homologized with traits of the primitively social bumble bees.

These findings emphasize the importance of the phylogenetic approach for investigating questions of behavioral evolution. This approach provides explicit hypotheses from which to test questions of character evolution. Furthermore, it can identify critical taxa for which comparative information is needed, thus paving the way for new areas of research. The development of a well-corroborated phylogenetic hypothesis for the Apidae should be a high priority for evolutionary studies of apid biology. There are some unique advantages that molecular data can bring to the study of apid phylogeny. A more complete understanding of apid diversity, however, will require further input from morphological and molecular systematists, behaviorists and ecologists, alike.

Acknowledgments

I thank Norma Gieg-Sinclair and Jim Whitfield for invaluable technical help; Bob DuBose and John Patton for advice, including preparation of oligonucleotides; David Roubik for providing the stingless bees for the study; Allan Larson, Alan Templeton, and Jim Whitfield for valuable discussions; and Byron Alexander, Allan Larson, Charles Michener, Gene Robinson, and Jim Whitfield for reading the manuscript. This research was supported in part by an NIH Postdoctoral Training Grant, an NIH BRSG grant to Allan Larson, and NIH grant 5RO1GM3157 to Alan Templeton.

Literature Cited

Aquadro, C. F. and B. D. Greenberg. 1983. Human mitochondrial DNA variation and evolution: Analysis of nucleotide sequences from seven individuals. *Genetics* 103:287-312.

Avise, J. C., J. Arnold, R. M. Ball, E. Bermingham, T. Lamb, J. E. Neigel, C. A. Reeb and N. C. Saunders. 1987. Intraspecific phylogeography: the mitochondrial DNA bridge between population genetics and systematics. *Annu. Rev. Ecol. Syst.* 8:489-522.

Baum, D. A. and A. Larson. 1991. Adaptation reviewed: A phylogenetic methodology for studying character macroevolution. *Syst. Zool.* (in press).

Brooks, D. R. and D. A. McLennan. 1991. *Phylogeny, Ecology and Behavior.* University of Chicago Press: Chicago.

Brown, W. M. 1985. The mitochondrial genome of animals. In *Molecular Evolutionary Genetics.* R. J. MacIntyre, ed. Plenum: NY, London.

Brown, W. M., E. M. Prager, A. Wang and A. C. Wilson. 1982. Mitochondrial DNA sequences of primates: Tempo and mode of evolution. *J. Mol. Evol.* 18:225-239.

Cameron, S. A. and G. E. Robinson. 1990. Juvenile hormone does not affect division of labor in bumble bee colonies (Hymenoptera: Apidae). *Ann. Entomol. Soc. Am.* 83:626-631.

Carpenter, J. M. 1989. Testing scenarios: wasp social behavior. *Cladistics* 5:131-144.

Clary, D. O. and D. R. Wolstenholme. 1985. The mitochondrial DNA molecule of *Drosophila yakuba*: Nucleotide sequence, gene organization, and genetic code. *J. Mol. Evol.* 22:252-271.

Clutton-Brock, T. and P. Harvey. 1984. Comparative approaches to investigating adaptation. In *Behavioral Ecology: An Evolutionary Approach*, 2nd Ed., J. R. Krebs and N. B. Davies, eds. Blackwell: Oxford.

Coddington, J. A. 1988. Cladistic tests of adaptational hypotheses. *Cladistics* 4:3-22.

De Queiroz, K. 1989. *Morphological and biochemical evolution in the sand lizards* (Appendix 3). Ph.D. dissertation. Univ. Calif., Berkeley.

Donoghue, M. J. 1989. Phylogenies and the analysis of evolutionary sequences, with examples from seed plants. *Evolution* 43:1137-1156.

Donoghue, M. J. and P. D. Cantino. 1984. The logic and limitations of the outgroup substitution approach to cladistic analysis. *Syst. Bot.* 9:192-202.

DuBose, R. F. and D. L. Hartl. 1990. Rapid purification of PCR products for DNA sequencing using Sepharose CL-6B spin columns. *Biotechniques* 8:271-274.

Dyer, F. C. 1987. New perspectives on the dance orientation of the asian honey bees. In *Neurobiology and Behavior of Honeybees.* R. Menzel and A. Mercer, eds. Springer-Verlag: Berlin, Heidelberg, New York.

Dyer, F. C. and T. D. Seeley. 1989. On the evolution of the dance language. *Am. Nat.* 133:580-590.

Farris, J. S. 1983. The logical basis of phylogenetic analysis. In *Advances in Cladistics* Vol 2: Proc. 2nd Meet. Willi Hennig Soc., N. I. Platnick and V. A. Funk, eds. Columbia University Press: New York.

------. 1988. Hennig 86: Version 1.5. Univ. of Stony Brook, NY.

Fauron, C. M-R. and D. R. Wolstenholme. 1980. Extensive diversity among *Drosophila* species with respect to nucleotide sequences within the adenine + thymine rich region of mitochondrial DNA molecules. *Nucleic Acids Res.* 11:2439-2453.

Felsenstein, J. 1981. Evolutionary trees from DNA sequences: A maximum likelihood approach. *J. Mol. Evol.* 17:368-376.

------. 1985a. Confidence limits on phylogenies with a molecular clock. *Syst. Zool.* 34:152-161.

------. 1985b. Confidence limits on phylogenies: An approach using the bootstrap. *Evolution* 39:783-791.

------. 1985c. Phylogenies and the comparative method. *Am. Nat.* 125:1-15.

------. 1988a. PHYLIP (Phylogeny Inference Package) Version 3.1 Manual.

------. 1988b. Phylogenies from molecular sequences: inference and reliability. *Annu. Rev. Genet.* 22:521-565.

Gerbi, S. A. 1985. Evolution of ribosomal DNA. In *Molecular Evolutionary Genetics.* R. J. MacIntyre, ed. Plenum: New York.

Goodman, M. 1989. Emerging alliance of phylogenetic systematics and molecular biology: A new age of exploration. In *The Hierarchy of Life.*, B. Fernholm, K. Bremer and H. Jörnwall, eds. Elsevier Science Publishers B. V. (Biomedical Division).

Gould, J. L., F. C. Dyer and W. F. Towne. 1985. Recent progress in the study of the dance language. *Fortschr. Zool.* 31:141-161.

Greenberg, B. D., J. E. Newbold and A. Sugino. 1983. Intraspecific nucleotide sequence variability surrounding the origin of replication in human mitochondrial DNA. *Gene* 21:33-49.

Greene, H. W. 1986. Diet and arboreality in the emerald monitor, *Varanus prasinus*, with comments on the study of adaptation. *Fieldiana Zool.* (NS) 31:1-12.

Gyllensten, U. B., D. Wharton and A. C. Wilson. 1985. Maternal inheritance of mitochondrial DNA during backcrossing of two species of mice. *J. Hered.* 76:321-324.

Harrison, R. G. 1989. Animal mitochondrial DNA as a genetic marker in population and evolutionary biology. *TREE* 4:6-11.

Hendy, M. D. and D. Penny. 1982. Branch and bound algorithms to determine minimal evolutionary trees. *Math. Biosci.* 59:277-290.

Hillis, D. M. and C. Moritz. 1990. *Molecular Systematics.* D. M. Hillis and C. Moritz, eds. Sinauer: Mass.

Hixon, J. E. and W. M. Brown. 1986. A comparison of the small ribosomal RNA genes from the mitochondrial DNA of the great apes and humans: Sequence, structure, evolution, and phylogenetic implications. *Mol. Biol. Evol.* 3:1-18.

Huey, R. B. 1987. Phylogeny, history and the comparative method. In *New Directions in Ecological Physiology*. M. E. Feder, A. F. Bennett, W. W. Burggren and R. B. Huey, eds. Cambridge University Press: Cambridge. Pp. 76-98.

Huey, R. B. and A. F. Bennett. 1987. Phylogenetic studies of coadaptation: Preferred temperatures versus optimal performance temperatures of lizards. *Evolution* 41:1098-1115.

Innis, M. A., D. H. Gelfand, J. J. Sninsky and T. J. White. 1990. *PCR Protocols. A Guide to Methods and Applications*. Academic Press: San Diego.

Jander, R. 1976. Grooming and pollen manipulation in bees (Apoidea): The nature and evolution of movements involving the foreleg. *Physiol. Entomol.* 1:179-194.

Kimsey, L. S. 1984. A re-evaluation of the phylogenetic relationships in the Apidae (Hymenoptera). *Syst. Entomol.* 9:435-441.

Koeniger, G. and N. Koeniger. 1990. Evolution of reproductive behavior in honey bees. In *Social Insects and the Environment*. B. Mallik, G. K. Veeresh and C. A. Viraktamath, eds. Proc. 11th Int. Cong. IUSSI. Oxford & IBH Publishing Co.: New Delhi, Bombay, Calcutta.

Lansman, R. A., J. C. Avise, C. F. Aquadro, J. F. Shapira and S. W. Daniel. 1983. Extensive genetic variation in mitochondrial DNAs among geographic populations of the deer mouse, *Peromyscus maniculatus. Evolution* 37:1-16.

Larson, A. 1984. Neontological inferences of evolutionary pattern and process in the salamander family Plethodontidae. *Evol. Biol.* 17:119-217.

Maddison, W. P. 1990. A method for testing the correlated evolution of two binary characters: Are gains or losses concentrated on certain branches of a phylogenetic tree? *Evolution* 44:539-557.

Maddison, W. P., M. J. Donoghue and D. R. Maddison. 1984. Outgroup analysis and parsimony. *Syst. Zool.* 33:83-103.

Maniatis, T., E. F. Fritsch and J. Sambrook. 1982. *Molecular Cloning: A Laboratory Manual.* Cold Spring Harbor Publications: Cold Spring Harbor, NY.

McLennan, D. A., D. R. Brooks and J. D. McPhail. 1988. The benefits of communication between comparative ethology and phylogenetic systematics: A case study using gasterosteid fishes. *Can. J. Zool.* 66:2177-2190.

Michener, C. D. 1944. Comparative external morphology, phylogeny, and classification of the bees (Hymenoptera). *Bull Am. Mus. Nat. Hist.* 82:151-326.

------. 1974. *The Social Behavior of the Bees*. Harvard University Press, Cambridge, Mass.

------. 1990a. Classification of the Apidae. *Univ. Kans. Sci. Bull.* 54:75-164.

------. 1990b. Caste in xylocopine bees. In *An Evolutionary Approach to Castes and Reproduction* (W. Engels, Ed.). Springer Verlag.

Mickevich, M. F. and S. J. Weller. 1990. Evolutionary character analysis: tracing character change on a cladogram. *Cladistics* 6:137-170.

Moritz, C., T. E. Dowling and W. M. Brown. 1987. Evolution of animal mitochondrial DNA: relevance for population biology and systematics. *Annu. Rev. Ecol. Syst.* 18:269-292.

Mullis, K. B. and F. Faloona. 1987. Specific synthesis of DNA in vitro via a polymerase catalyzed chain reaction. *Methods Enzymol.* 155:335-350.

Mullis, K. B., F. Faloona, S. J. Scharf, R. K. Saiki, G. T. Horn and H. A. Erlich. 1986. Specific enzymatic amplification of DNA in vitro: The polymerase chain reaction. *Cold Spring Harbor Symposium on Quantitative Biology.* 51:263-273.

Page, R. E., Jr. 1980. The evolution of multiple mating behavior by honey bee queens (*Apis mellifera* L.) *Genetics* 96:263-273.

Pagel, M. D. and P. H. Harvey. 1988. Recent developments in the analysis of comparative data. *Q. Rev. Biol.* 63:413-440.

Plant, J. D. and H. F. Paulus. 1987. Comparative morphology of the postmentum of bees (Hymenoptera:Apoidea) with special remarks on the evolution of the lorum. *Z. zool. Syst. Evolutionforsch.* 25:81-103.

Ridley, M. 1983. *The Explanation of Organic Diversity. The Comparative Method and Adaptations of Mating.* Clarendon: Oxford.

Remane, A. 1952. Die Grundlagen des natürlichen Systems der vergleichenden Anatomie und der Phylogenetik. 2. Geest und Portig K. G., Leipzig.

Robinson, G. E. 1985. Effects of juvenile hormone analogue on honey bee foraging behaviour and alarm pheromone production. *J. Insect Physiol.* 31:277-282.

------. 1987. Regulation of honey bee age polyethism by juvenile hormone. *Behav. Ecol. Sociobiol.* 20:329-338.

Saiki, R. K., S. Scharf, F. Faloona, K. B. Mullis, G. T. Horn, H. A. Erlich and N. Arnheim. 1985. Enzymatic amplification of ß-globin genomic sequences and restriction site analysis for diagnosis of sickle cell anemia. *Science* 230:1350-1354.

Sakagami, S. F. and C. D. Michener. 1987. Tribes of Xylocopinae and origin of the Apidae (Hymenoptera: Apoidea). *Ann. Entomol. Soc. Am.* 80:439-450.

Sanderson, M. J. 1989. Confidence limits on phylogenies: the bootstrap revisited. *Cladistics* 5:113-129.

Sanger, F., S. Nicklen and A. R. Coulson. 1977. DNA sequencing with chain-terminating inhibitors. *Proc. Natl. Acad. Sci. USA* 74:5463-5467.

Satta, Y., H. Ishiwa and S. I. Chigusa. 1987. Analysis of nucleotide substitutions of mitochondrial DNAs in *Drosophila melanogaster* and its sibling species. *Mol. Biol. Evol.* 4:638-650.

Schmitz, J. and R. F. A. Moritz. 1990. Mitochondrial DNA variation in social wasps (Hymenoptera, Vespidae). *Experientia* 46:1068-1072.

Seeley, T. D. 1985. *Honeybee Ecology: A Study of Adaptation in Social Life.* Princeton University Press: Princeton NJ.

Sheppard, W. S. and S. H. Berlocher. 1989. Allozyme variation and differentiation among *Apis* species. *Apidologie* 20:419-431.

Sillén-Tullberg, B. 1988. Evolution of gregariousness in aposematic butterfly larvae: A phylogenetic analysis. *Evolution* 42:293-305.

Simon, C., S. Pääbo, T. Kocher and A. C. Wilson. 1990. Evolution of mitochondrial ribosomal RNA in insects as shown by the polymerase chain reaction. In *Molecular Evolution.* M. T. Clegg and S. J. O'Brien, eds., UCLA Symposium on Molecular and Cellular Biology, New Series, vol. 122. Alan R. Liss, Inc.: New York.

Simon, C., A. Franke and A. Martin. 1991. The polymerase chain reaction: DNA extraction and amplification. In *Molecular Taxonomy.* G. M. Hewitt, ed. NATO Advanced Studies Institute, Springer Verlag: Berlin.

Smith, D. R. 1991. African bees in the Americas: insights from biogeography and genetics. *TREE* 6:17-21.

Smith, D. R. and W. M. Brown. 1990. Restriction endonuclease cleavage site and length polymorphisms in mitochondrial DNA of *Apis mellifera mellifera* and *A. m. carnica* (Hymenoptera: Apidae). *Ann. Entomol. Soc. Am.* 83:81-88.

Smith, D. R., M. F. Palopoli, B. R. Taylor, L. Garnery, J.-M. Cornuet, M. Solignac, and W. M. Brown. 1991. Geographic overlap of two classes of mitochondrial DNA in Spanish honey bees (*Apis mellifera iberica*). *J. Hered.* 82:96-100.

Swofford, D. L. 1990. PAUP: Phylogenetic Analysis Using Parsimony, Version 3.0L. Illinois Natural Hist. Surv., Champaign, IL.

Swofford, D. L. and W. P. Maddison. 1987. Reconstructing ancestral character states under Wagner parsimony. *Math. Biosci.* 87:199-229.

Templeton, A. R. 1983. Convergent evolution and non-parametric inferences from restriction fragment and DNA sequence data. In *Statistical Analysis of DNA Sequence Data.*, B. Weir, ed. Marcel Dekker: New York. Pp. 151-179.

------. 1987. Nonparametric inference from restriction cleavage sites. *Mol. Biol. Evol.* 4:315-319.

Watrous, L. E. and Q. D. Wheeler. 1981. The outgroup comparison method of character analysis. *Syst. Zool.* 30:1-11.

Wheeler, W. C. 1989. The systematics of insect ribosomal DNA. In *The Hierarchy of Life.* B. Fernholm, K. Bremer and H. Jörnwall, eds. Elsevier Science Publishers B. V. (Biomedical Division).

Winston, M. L. 1987. *The Biology of the Honey Bee.* Harvard University Press: Cambridge MA.

Winston, M. L. and C. D. Michener. 1977. Dual origin of highly social behavior among bees. *Proc. Natl. Acad. Sci. USA* 74:1135-1137.

Wolstenholme, D. R. and D. O. Clary. 1985. Sequence evolution of *Drosophila* mitochondrial DNA. *Genetics* 109:725-744.

5

Ribosomal DNA Diversity in Apidae

Walter S. Sheppard and Bruce A. McPheron

The evolution of bees belonging to the family Apidae is a topic of interest to evolutionary biologists for a number of reasons. First, the attainment of eusociality in some, but not all, of the species is itself suggestive of a special role the group may ultimately play in disclosing the evolution of this remarkable phenomenon. Ants and termites, while themselves representing pinnacles of eusocial development, are groups whose non-social lineages have long been extinct. However, within the Apidae we can posit questions that may be solvable through comparative analysis of variation present in extant species. For example, did eusociality arise in the Apidae but once, or independently on two or more occasions? Similarly, hypotheses pertaining to the evolution of specific behaviors, morphological and physiological adaptations, and reproductive strategies may be testable within the family. Within the Apinae (consisting of the small genus *Apis*), for example, the recent discovery of a fundamentally different form of wintering behavior in *Apis laboriosa* (Underwood, 1990) and a continuing controversy over cavity-nesting as primitive or derived (Koeniger, 1976; Sakagami et al., 1980; Ruttner, 1988) obscures phylogenetic estimates.

Most attempts to understand the evolution of Apidae, from the generic level to familial, have been based on the assessment of comparative morphology or behavior. Discussion of character polarity and convergence and their implications for phylogeny estimation in this group still continues (Chapters 1, 3 and 4, this volume), and views of evolution within Apidae necessarily reflect such conclusions (Michener, 1974; Winston and Michener, 1977; Kerr, 1987).

Recently, molecular characters have become available for the study of evolution in numerous taxa. One subset of these characters includes ribosomal DNA (rDNA) sequences used for phylogeny estimation in insects (Vossbrinck and Friedman, 1989; Wheeler, 1989; McPheron et al., unpublished). While these insect studies have been directed toward investigating relationships at the familial or ordinal level, the potential for rDNA sequence analysis to resolve relationships among recently diverged taxa has been discussed (Coen et al., 1982; Lassner et al., 1987; Smith, 1989). In fact, one of the intriguing aspects of sequence data is that sequences often range from highly variable to highly conserved among taxa, depending on the particular molecules and regions being analyzed. This property is shared with morphological and behavioral data and indicates that molecular data, also, may be used to study evolutionary questions at multiple systematic levels. As a prelude to further study of the group, we present here preliminary molecular data on evolutionary questions within the Apidae and Apinae (*Apis*) derived from ribosomal DNA. The techniques used for this project are direct sequencing of the small and large subunits of ribosomal RNA, and amplification (via the polymerase chain reaction) and sequencing of mitochondrial and nuclear ribosomal DNA. Each molecule and technique will be briefly discussed, along with the evolutionary hypotheses relevant to the Apidae that may be testable by each. Although the evolutionary assertions that we can make at this juncture are necessarily limited, we believe that the potential benefit of such a multilevel approach to evolutionary systematic study of the group will be evident.

Collection of Bee Material

All bees were collected as adults, killed on dry ice or in liquid nitrogen, and subsequently stored at -80°C. Due to the preliminary nature of this report, we will not list details of the collecting localities; this information is available from the senior author.

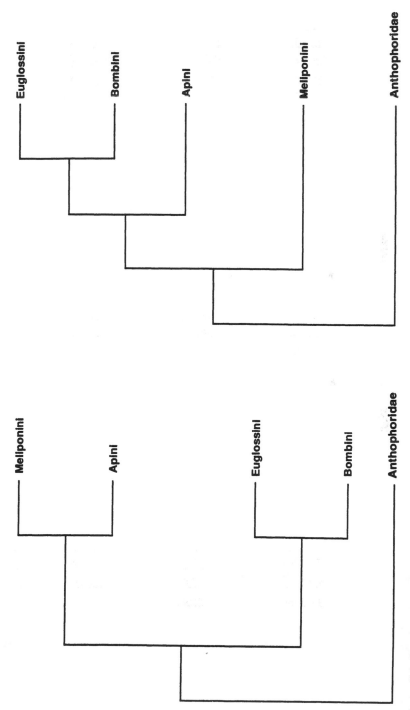

Figure 5.1 Proposed phylogenies of tribes in the Apidae (after Michener, 1974; Winston and Michener, 1977): a) Single evolution of eusocial behavior; b) Dual evolution of eusocial behavior.

Table 5.1 Primer sequences used in reverse transcriptase-mediated primer extension of ribosomal RNA (rRNA).

Primer[1]	Sequence	Position[2]
18H	5'- TCAATTCCTTTAAGTTTCAGC - 3'	1215-1235
580	5' - GGTCCGTGTTTCAAGACGG - 3'	4049-4067

[1]primer 18H from Hamby and Zimmer (1988), primer 580 from Vossbrinck and Friedman (1989). 18H is in the small subunit rRNA gene and 580 is in the large subunit rRNA gene. Primer sequences are reverse complement to published sequence.
[2]position based on numbering system of Tautz et al. (1988) for *Drosophila melanogaster*.

Nucleic Acid Extraction

All analyses were performed on nucleic acid extractions from individual bees using the following protocol (modified from Coen et al., 1982). A single bee thorax was placed in 1 ml of cold Solution 1 (10 mM Tris-HCl [pH 8.0], 60 mM NaCl, 10 mM Na_2 EDTA [pH 8.0], 5% sucrose) in a 15 ml Corex centrifuge tube kept on ice. Solution 2 (300 mM Tris-HCl [pH 8.0], 20 mM Na_2 EDTA, 1.25% SDS, 5% sucrose) was prepared by adding 12 µl diethyl pyrocarbonate (DEPC) per 1 ml of Solution 2. Solution 2 plus DEPC was left on a stirring platform with a magnetic stir bar spinning until needed to facilitate dispersion of the DEPC in the solution. The insect tissue was homogenized with a Teflon pestle on ice just until tissue structure was disrupted, and 1 ml of Solution 2 was added immediately. This mixture was placed on ice for 15 minutes.

An equal volume of buffer-equilibrated phenol was added, vortexed, and placed on ice for 3 minutes. Centrifugation for 5 minutes at 3000 x *g* at 4°C separated the layers. The upper, aqueous layer was removed with a cut pipet tip or cell saver tip to a clean 15 ml Corex tube. This extraction procedure was repeated twice, first with phenol:chloroform:isoamyl alcohol (25:24:1), and second with chloroform:isoamyl alcohol (24:1). After the final extraction, the upper, aqueous layer was transferred with a cut pipet tip or cell saver tip to a siliconized 15 ml Corex tube. An equal volume of modified 1X TE (10 mM Tris-HCl, 0.1 mM Na_2 EDTA, pH 8.0) was added. This modified TE recipe (lower Na_2 EDTA concentration) is important if the purified nucleic acids are to be used for amplification with *Taq* polymerase, which is highly dependent upon $[Mg^{2+}]$. The solution was made 0.3 M with respect to sodium acetate and 2.5 volumes of cold absolute ethanol were added. The tube was covered with parafilm, inverted 10 times, and placed at -20°C for at least 2 hours to precipitate nucleic acids.

Centrifugation at 12,000 x *g* for 30 minutes at 4°C pelleted the nucleic acid. This pellet was resuspended in 1 ml modified 1X TE, and a second ethanol precipitation was performed. The precipitate was pelleted by another high speed centrifugation, and the pellet was dried under vacuum. When thoroughly dry, the pellet was resuspended in 250 μl modified TE and stored at -20°C until use.

Direct Sequencing of
Ribosomal RNA

This work began out of an effort to find genetic markers suitable for estimating the phylogeny of the genus *Apis*. Isozyme (allozyme) studies of *Apis* (Sheppard and Berlocher, 1989; Chapter 7, this volume) have revealed considerable electrophoretic variability, too much, in fact, to clearly support any of the proposed phylogenies. In contrast, phylogenetic studies based on comparisons of ribosomal RNA have shown certain regions of the molecule to be relatively conserved and suitable for the study of higher taxonomic questions (Vossbrinck and Friedman, 1989; McPheron et al., unpublished). Although the regions of the small and large subunit rRNAs we sequenced proved uninformative for the question of *Apis* phylogeny due to a paucity of variation (Sheppard, unpublished), the possibility remained to use sequence analysis of these relatively conserved molecules to estimate apid phylogeny.

Of the questions often discussed in the literature concerning apid evolution, the dual or single origin for eusociality is perhaps most volatile (Michener, 1944, 1974; Winston and Michener, 1977; Kerr, 1987; Chapters 3 and 4, this volume; Figure 5.1). This question can be rephrased

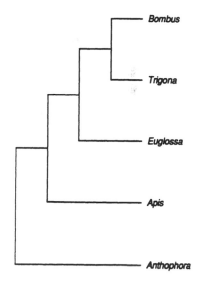

Figure 5.2 Single most parsimonious tree generated by PAUP 3.0h using branch and bound algorithm on 7 phylogenetically informative rRNA characters; tree length = 11 steps; consistency index = 0.82.

Table 5.2 Matrix of phylogenetically informative ribosomal RNA sites from primer 580.

Taxon:

Anthophora abrupta	CACATCA
Apis mellifera	CACACTG
Euglossa imperialis	CAGGTTG
Bombus terricola	TGAGCGA
Trigona capitata	TGAGTGG

within the context of phylogenetic systematics by asking "are the Apini and Meliponini sister taxa?" If they are, then a single origin for eusociality is most parsimonious and, if they are not, then a dual origin for sociality becomes most likely, since the sister taxon for one of them would include non-social members. A recent reanalysis and interpretation of morphological characters supports the single origin hypothesis (Chapter 3, this volume). In an attempt to bring molecular data to bear on the question of relationships within the Apidae, we sequenced portions of the 18S and 28S ribosomal RNAs for five taxa within the group: *Anthophora abrupta, Apis mellifera, Euglossa imperialis, Bombus terricola,* and *Trigona capitata.*

We followed the protocol of Hamby and Zimmer (1988) for direct sequencing of both small and large subunit ribosomal RNAs (rRNA) using reverse transcriptase-mediated primer extension. Total nucleic acid was quantified spectrophotometrically, and approximately 6 μg were used for each sequencing reaction. Four *p*moles of an oligonucleotide primer (Table 5.1) were annealed to the rRNA sample. The reaction mixture was divided equally into four tubes, each containing all four deoxynucleotides, including ^{32}P-labeled dCTP (650 Ci/mmole, ICN Radiochemicals, Irvine, CA), and one dideoxynucleotide. Chain extension occurred in the presence of 4.5 units (U) of reverse transcriptase (16.6 U/μl, avian myeloblastosis virus, Life Sciences, Inc., St. Petersburg, FL) during an incubation of 10 minutes at 42°C followed by 10 minutes at 50°C. A chase step (an excess of unlabeled deoxynucleotides) of 10 minutes at 50°C followed by 10 minutes at 60°C permitted complete extension of any chains that had not terminated due to addition of a dideoxynucleotide. The reaction was stopped by the addition of formamide dye solution. The complementary DNA strands generated in these sequencing reactions were resolved on a 9% sequencing gel run at 1800 V. The gel was dried onto Whatman 3MM chromatography paper on a gel dryer, and an autoradiograph was produced by exposure of Kodak X-Omat AR film to the dried gel for 8 - 24 hours at room temperature.

Alignment of ribosomal RNA sequence data was performed with the alignment algorithm available in the Intelligenetics computer package. Each sequence was aligned separately to *Bombus terricola* sequence. Final editing was by eye on a word processor. Only those positions representing unambiguous mutational events were included in the matrix. Certain short regions of sequence generated with each primer differed in length and were not alignable, even though sequences on either side of these regions were highly conserved among taxa. These regions are not included in the analysis because alignment would be highly subjective. Autapomorphies, characters distinguishing a

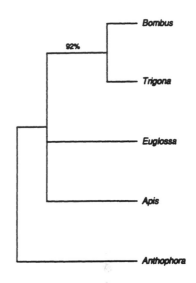

Figure 5.3 Bootstrap 80% majority-rule consensus tree of rRNA data generated by PAUP 3.0h, branch and bound algorithm, bootstrap with 100 replications; tree length = 12 steps; consistency index = 0.75.

single taxon from all others, were also excluded from the analysis. Alignment of rRNA sequences among the five taxa yielded 102 positions for primer 18H and 135 positions for primer 580. Of these 237 characters just 7, all in the sequence resulting from primer 580, were phylogenetically informative (Table 5.2).

This matrix of phylogenetically informative sites was analyzed by PAUP version 3.0h (Swofford, 1990), using the branch and bound algorithm and the bootstrap option (100 replications). The single most parsimonious tree unites *Trigona* (Meliponini) and *Bombus* (Bombini) and makes *Euglossa* (Euglossini) the sister taxon to this clade (Figure 5.2). Bootstrap analysis of this data set shows 92% support for the *Trigona/Bombus* sister relationship and fails to resolve the position of the other apids (Figure 5.3). The relationship hypothesized between Bombini and Meliponini is not generally supported by morphological and behavioral data (Michener, 1974; Winston and Michener, 1977; Chapter 3, this volume), although Kerr (1987) suggested a sister group relationship for these taxa based on a shared physiological mechanism of caste determination. A similar classification is also supported by sequence data from mitochondrial rDNA (Chapter 4, this volume). Further work within this group continues in our

Table 5.3 Primer pairs used for PCR amplification of bee DNA.

Gene[1]	Primer	Sequence	Position[2]
ITS1	1975F	5'- TAACAAGGTTCCGTAGGTG - 3'	1955-1975
	35R	5'- AGCTRGCTGCGTTCTTCATCGA - 3'	2754-2775
12S	L1091	5'- AAAAAGCTTCAAACTGGGATTAGATACCCCACTAT - 3'	14588-14622
	H1478	5'- TGACTGCAGAGGGTGACGGGCGGTGTGT - 3'	14206-14233

[1] primers for ITS1, the internal transcribed spacer between nuclear 18S and 5.8S genes, were developed by Hugh Robertson (1975F) and Carl Woese (35R), University of Illinois at Urbana-Champaign; primers for mitochondrial 12S ribosomal RNA gene from Kocher et al. (1989)
[2] position for ITS1 primers based on numbering system of Tautz et al. (1988) for *Drosophila melanogaster*; position for 12S primers based on numbering system of Clary and Wolstenholme (1985) for *Drosophila yakuba*

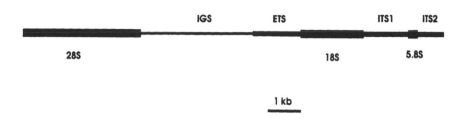

Figure 5.4 Schematic representation of an insect ribosomal DNA repeat unit; IGS (intergenic spacer), ETS (external transcribed spacer), ITS (internal transcribed spacer).

and other laboratories, and the inclusion of additional taxa and sequence data from these and other molecules should help resolve the phylogeny.

Polymerase Chain Reaction

In addition to the regions of rDNA known to code for functional sub-units (5.8S, 18S, 28S) in insects, there are several spacer regions located between the functional genes (Figure 5.4). One of these, the large intergenic spacer (IGS), occurs between the 28S and 18S genes of consecutive transcription units and has been shown to be highly variable among closely-related taxa (Collins et al., 1987; Lassner et al., 1987; Tautz et al., 1987), presumably due to lack of strong selection compared to functional genes (Coen et al., 1982) or the evolutionary processes within this multigene family (Tautz et al., 1987). The IGS may exhibit length variation within a species and within an individual (Beach et al., 1989; Collins et al., 1989). The internal transcribed spacers (ITS) that occur on either side of the 5.8S gene may also be relatively free from selection, and we feel that sequences from these regions may be systematically useful for more recently diverged bee taxa.

Mitochondrial genes are another source of phylogenetically useful sequence data. The 12S (small subunit) ribosomal RNA gene contains certain sequences that are relatively conserved throughout a wide range of animals (Kocher et al., 1989; Simon et al., 1990). Variation in DNA sequences between these conserved regions appears to be useful for more distantly-related taxa. Simon et al. (1990) illustrate some sequence conservation in Domain III of this molecule between *Magicicada* and *Drosophila*. Lacking sequence data from other insect taxa for comparison, we expected that portions of this molecule have accumulated sufficient sequence divergence to make it a reasonable candidate for study of apid phylogeny, since the Meliponini were distinct from the other apid tribes at least 40-50 million years before present (Michener, 1974; Ruttner, 1988).

We used the polymerase chain reaction (PCR) to amplify two regions of bee DNA: the internal transcribed spacer between the nuclear 18S and 5.8S ribosomal RNA genes (ITS1) and a section of the 12S ribosomal RNA gene from the mitochondrial DNA (12S). Primer sequences are presented in Table 5.3. PCR was conducted using GeneAmp^{TM} reagents (Perkin Elmer Cetus, Norwalk, CT). Reactions were carried out in 50 µl volumes, using 1 µl of the total nucleic acid extractions as template. For initial symmetrical amplifications, both primers for the gene to be amplified were present at 1 µM concentrations in the presence of 1.5 U/reaction of AmpliTaq^{TM} DNA polymerase (Perkin Elmer Cetus, Norwalk, CT). Amplification continued for 35 cycles with a denaturation step at 93°C for 1 minute, annealing at 50°C for 1 minute, and extension at 72°C for 2 minutes.

Amplified DNA (10 µl) was separated electrophoretically on 0.8% agarose gels (TBE buffer) and visualized with a hand-held UV light source following ethidium bromide staining. Small wells were cut in front of the bands, and these wells were filled with 1% low-melting-point agarose (NuSieve, FMC) in TBE buffer. When these plugs solidified, the bands were run into the plugs and electrophoresis was stopped. The center of this band was cut from the gel using a sterile scalpel. This small amount of agarose containing only amplified DNA was melted in 1 ml of water at 95°C.

Asymmetrical amplifications (Kocher et al., 1989) were conducted to prepare single-stranded ITS1 DNA suitable for sequencing. Reaction volumes were again 50 µl, using 1 µl of the gel purified double-stranded DNA as template. Primer 1975F was added at 1 µM, and primer 35R was present at 0.01 µM. Amplification continued for 40 cycles with a denaturation step at 93°C for 1 minute, annealing at 50°C for 1 minute, and extension at 72°C for 2 minutes. 10 µl of each reaction was run on 0.8% agarose gels (TBE buffer) to assay for amplification. The remaining 40 µl were placed in a Millipore 30,000 NMWL filter unit (Millipore Corporation, Bedford, MA), and the volume was increased to 300 µl with water. Samples were spun down (6000 x *g* at room temperature) to a volume of approximately 50 µl. Filtration was repeated two additional cycles. In the final cycle, the samples were spun to a volume of 10 µl, which was stored at 4°C until needed for sequencing. The appropriate sequencing primer for this product was 35R, the limiting primer in the asymmetrical PCR.

We were unable to consistently generate single-stranded copies of amplified 12S DNA via asymmetrical amplification, so we used the protocol developed by Higuchi and Ochman (1989) to produce single stranded DNA through a symmetrical PCR reaction. One primer (L1091) was treated with T4 polynucleotide kinase (Bethesda Research Labs, Bethesda, MD) to transfer a phosphate group to the 5'-hydroxyl terminus of the primer sequence by the forward reaction (Ausubel et al., 1987). This primer was used at the normal concentration (1 µM in final reaction) in a symmetrical PCR as described previously.

The double-stranded product was filtered three times through a Millipore 30,000 NMWL filter unit at 6000 x *g*; the final volume of the sample was reduced to 43 µl. The 5'-phosphoryl strand of the duplex was then digested away by the action of lambda exonuclease (Bethesda Research Labs, Bethesda, MD) (Ausubel et al., 1987), leaving a single-stranded product. This product was filtered three times through a Millipore 100,000 NMWL filter unit at 6000 x *g*, retaining a final sample volume of 10 µl after the third spin. This single stranded DNA corresponds to the strand that was extended from primer H1478 (the non-phosphorylated primer). The appropriate sequencing primer for this strand was L1091.

Sequencing of Amplified DNA

Dideoxy sequencing of amplified DNA was performed using the Sequenase[R] reagent kit (United States Biochemical Corporation, Cleveland, OH) and a modification of their sequencing protocol. 7 µl of the amplification product, 2 µl of the 5X Sequenase[R] annealing buffer, and 1 µl of a 20 µM stock of the appropriate primer were heated to 95°C for 6 minutes and cooled at room temperature for 1 minute. 5.5 µl of the labeling reaction mix made according to the Sequenase[R] protocol (dithiothreitol, dilute Sequenase[R] labeling mix, diluted Sequenase[R], and the appropriate nucleotide [^{32}P-labeled dATP (650 Ci/mmole, ICN Radiochemicals, Irvine, CA) for ITS1 or ^{35}S-labeled dATP (1500 Ci/mmole, DuPont NEN, Boston, MA) for 12S)] were added to each annealing tube, which were immediately placed onto ice for 3 minutes. An aliquot of 3.5 µl from this labeling tube was added to each of four termination tubes (each containing a different dideoxynucleotide), and this was incubated at 37°C for 3 minutes. The reactions were stopped by the addition of formamide dye solution. DNAs were separated on a 9% sequencing gel, run at 1800 V. The gel was dried onto Whatman 3MM chromatography paper on a gel dryer, and an autoradiograph was produced by exposure of Kodak X-Omat AR film to the dried gel for 18 - 36 hours at room temperature.

An amplification product of approximately 400 base pairs (bp) was consistently obtained using the ITS1 primers with nucleic acid extractions from *Anthophora abrupta*, two Old World races of *Apis mellifera* (*A. m. intermissa* and *A. m. ligustica*), and *A. mellifera* from Argentina (from a sample with mitochondrial DNA and morphological characters similar to African *A. m. scutellata*). While the size of this fragment is at variance with that of *Drosophila melanogaster* (Tautz et al., 1988), its consistent occurence in the bee taxa investigated, absence in amplified Symphyta (Tenthredinidae: *Euura s-nodus*) and Diptera (Tephritidae: *Ceratitis capitata*) DNA, and absence in negative controls indicate a potential degree of specificity. Sequences have only been obtained from *Anthophora* (83 bp) and the Argentine *A. mellifera* (132 bp), so the utility of this region for phylogenetic study in the group cannot be determined. However, alignment of the *Anthophora* and *Apis* sequences yields only 64% base pair matching within a short region with nearly complete sequence similarity at either end. We speculate that the ITS1 region may be better suited for comparisons of more recently diverged bee taxa, such as *Apis* species or races.

The 12S primers consistently yielded an amplification product of approximately 360 bp in *Colletes thoracicus*, *Anthophora abrupta*, *Euglossa imperialis*, *Bombus terricola*, *Apis mellifera* (Argentinian sample), *A. cerana*, *A. florea*, and *A. dorsata*. Sequence data obtained from this amplification indicate surprising degrees of divergence. *A. dorsata* (208 bp) and *A. mellifera* (233 bp) exhibit only 79% sequence identity. Further sequence comparisons are required to verify that this region of the 12S mitochondrial rDNA is indeed diverging at this rate. This rapid evolution would make it valuable for study within *Apis*.

Conclusions

Hypotheses about evolution and systematics of the Apoidea and Apidae are based largely on information accumulated from the study of morphology and behavior. We can turn to Aristotle's description of racial variation within *Apis mellifera* (Natural History, 344-342 B.C.) as an example of early recognition that morphological characters are indicative of underlying genetic variation. As molecular systematic techniques continue to provide additional genetically-based characters, suitable for phylogenetic analysis in extant and even extinct taxa, we can expect continued testing of these hypotheses. Our preliminary results indicate that variation within ribosomal DNA is suitable to provide characters for the study of the Apidae and, perhaps, related families. An eventual synthesis of information from both molecular and non-molecular methodologies may be our best hope to significantly improve understanding of bee evolution.

Acknowledgments

We thank the following for help in collecting taxa used in this research: Hilary de Alwis, Stephen Hight, David Roubik, Carol Sheppard, and Gary Steck. A portion of this research was funded by a Pennsylvania State University Research Initiation Grant to BAM. Mention of a trademark, proprietary product, or vendor does not constitute a guarantee or warranty by the USDA and does not imply its approval to the exclusion of other products or vendors that may also be suitable.

Literature Cited

Ausubel, F. M., R. Brent, R. E. Kingston, D. D. Moore, J. G. Seidman, J. A. Smith and K. Struhl (eds.). 1987. *Current Protocols in Molecular Biology, Volume 1.* John Wiley and Sons: New York.

Beach, R. F., D. Mills and F. H. Collins. 1989. Structure of ribosomal DNA in *Anopheles albimanus* (Diptera: Culicidae). *Ann. Entomol. Soc. Am.* 82:641-648.

Clary, D. O. and D. R. Wolstenholme. 1985. The mitochondrial DNA molecule of *Drosophila yakuba*: nucleotide sequence, gene organization, and genetic code. *J. Mol. Evol.* 22:252-271.

Coen, E. S., J. M. Thoday and G. Dover. 1982. Rate of turnover of structural variants in the rDNA gene family of *Drosophila melanogaster. Nature* 295:564-568.

Collins, F. H., M. A. Mendez, M. O. Rasmussen, P. C. Mehaffey, N. J. Besansky and V. Finnerty. 1987. A ribosomal RNA gene probe differentiates member species of the *Anopheles gambiae* complex. *Am. J. Trop. Med. Hyg.* 37:37-41.

Collins, F. H., S. M. Paskewitz and V. Finnerty. 1989. Ribosomal RNA genes of the *Anopheles gambiae* species complex. In *Advances in Disease Vector Research, Vol. 6.*, K. F. Harris, ed. Springer-Verlag: New York. Pp. 1-28.

Hamby, R. K. and E. A. Zimmer. 1988. Ribosomal RNA sequences for inferring phylogeny within the grass family (Poaceae). *Plant Syst. Evol.* 160:29-37.

Higuchi, R. G. and H. Ochman. 1989. Production of single-stranded DNA templates by exonuclease digestion following the polymerase chain reaction. *Nucleic Acids Res.* 17:5865.

Kerr, W. E. 1987. Sex determination in bees. XVII. Systems of caste determination in the Apinae, Meliponinae and Bombinae and their phylogenetic implications. *Rev. Bras. Genet.* 10:685-694.

Koeniger, N. 1976. Neue Aspekte der Phylogenie innerhalb der Gattung *Apis. Apidologie* 7:357-366.

Kocher, T. D., W. K. Thomas, A. Meyer, S. V. Edwards, S. Pääbo, F. X. Villablanca and A. C. Wilson. 1989. Dynamics of mitochondrial DNA evolution in animals: amplification and sequencing with conserved primers. *Proc. Natl. Acad. Sci., USA* 86:6196-6200.

Lassner, M., O. Anderson and J. Dvorak. 1987. Hypervariation associated with a 12-nucleotide direct repeat and inferences on intergenomic homogenization of ribosomal RNA gene spacers based on the DNA of a clone from wheat *Nor-D3* locus. *Genome* 29:770-781.

Michener, C. D. 1944. Comparative external morphology, phylogeny and classification of the bees (Hymenoptera). *Bull. Am. Mus. Nat. Hist.* 82:151-326.

------. 1974. *The Social Behavior of the Bees.* Harvard University Press: Cambridge, Mass.

Ruttner, F. 1988. *Biogeography and Taxonomy of Honeybees.* Springer-Verlag: Heidelberg, New York.

Sakagami, S. F., T. Matsumura and K. Ito. 1980. *Apis laboriosa* in Himalaya, the little known world largest honey bee (Hymenoptera: Apidae). *Insecta Matsumurana* 19:47-77.

Sheppard, W. S. and S. H. Berlocher. 1989. Allozyme variation and differentiation among four *Apis* species. *Apidologie* 20:419-430.

Simon, C., S. Pääbo, T. D. Kocher and A. C. Wilson. 1990. Evolution of mitochondrial ribosomal RNA in insects as shown by the polymerase chain reaction. In *Molecular Evolution.* M. Clegg and S. O'Brien, eds. Alan R. Liss, Inc.: New York pp. 235-244

Smith, A. B. 1989. RNA sequence data in phylogenetic reconstruction: testing the limits of its resolution. *Cladistics* 5:321-344.

Swofford, D. L. 1990. PAUP, phylogenetic analysis using parsimony, version 3.0, users manual. Illinois Natural History Survey, Champaign, IL.

Tautz, D., C. Tautz, D. Webb and G. A. Dover. 1987. Evolutionary divergence of promoters and spacers in the rDNA family of four *Drosophila* species: implications for molecular coevolution in multigene families. *J. Mol. Biol.* 195:525-542.

Tautz, D., J. M. Hancock, D. A. Webb, C. Tautz and G. A. Dover. 1988. Complete sequences of the rRNA genes of *Drosophila melanogaster. Mol. Biol. Evol.* 5:366-376.

Underwood, B. A. 1990. *The behavior and energetics of high altitude survival by the himalayan honey bee.* PhD dissertation. Cornell University, Ithaca, NY.

Vossbrinck, C. R. and S. Friedman. 1989. A 28s ribosomal RNA phylogeny of certain cyclorrhaphous Diptera based upon a hypervariable region. *Syst. Entomol.* 14:417-431.

Wheeler, W. C. 1989. The systematics of insect ribosomal DNA. In *The Hierarchy of Life.* B. Fernholm, K. Bremer and H. Jornvall, eds. Elsevier Science Publ. Pp. 307-321.

Winston, M. L. and C. D. Michener. 1977. Dual origin of highly social behavior among bees. *Proc. Natl. Acad. Sci. USA* 74:1135-1137.

6

Genetic Diversity in *Apis mellifera*

Jean-Marie Cornuet and Lionel Garnery

Apis mellifera is a widespread species that has differentiated into numerous geographic races or subspecies. These subspecies differ in various characteristics such as morphology, behavior, ecology, sensitivity to diseases and biochemical components. Because most of these characteristics have a genetic basis, the level of genetic diversity within the species as a whole may be considered high.

In order to evaluate and describe more precisely this genetic diversity, different methods have been used. The first was morphometrics, which deals with quantitative characters with generally high heritability. Nonetheless, this method, in which genotypes cannot be established through phenotypes, gives only an indirect and possibly biased measure of genetic diversity. The second approach has been the study of allozymic variation which, most of the time, allows the determination of genotypes. Finally, in the last four years, studies of the genetic variation of *Apis mellifera* at the DNA level have been developed.

In the last 15 years we have accumulated data on the variation between and within subspecies with these methods. With these results and those reported in the literature, we wish to compare the information provided by these three methods on the genetic structure of *Apis mellifera*.

Morphometry

Our first studies done in Monfavet (Station de Zoologie et d'Apidologie, Montfavet, France) used only 6 morphometric characters. This is very few compared to the 40 characters measured by Ruttner's

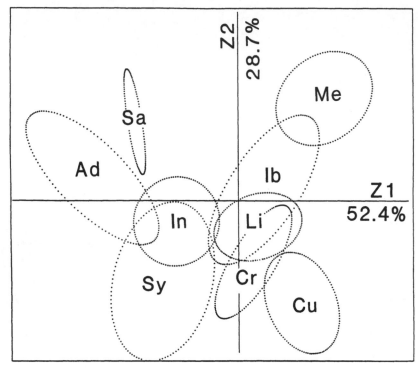

Figure 6.1 Factorial Discriminant Analysis of nine *Apis mellifera* subspecies based
on five morphological characters. Ad = *A. m. adansoni*, Cr = *A. m. carnica*, Cu =
A. m. caucasica, Ib = *A. m. iberica*, In = *A. m. intermissa*, Li = *A. m. ligustica*, Me
= *A. m. mellifera*, Sa = *A. m. sahariensis*, and Sy = *A. m. syriaca*.

team (e.g., Ruttner et al., 1978; Ruttner, 1988). However these data, in
association with performant multidimensional analysis (principal component analysis, factorial discriminant analysis) provided good
discrimination among subspecies and fair discrimination among
populations of the same subspecies.

Figure 6.1 gives an example of the kind of results that can be obtained
with this method. It shows ellipses of confidence (0.95 level) of 9
subspecies on the first plane of a factorial discriminant analysis. The
overlapping ellipses can be differentiated in the other planes. Morphometrics can be a very powerful tool for the discrimination of
populations. Two populations of the same subspecies located 8 km apart
and separated by a small ridge were partially discriminated: 80% of the
measured colonies could be correctly assigned to their original population
(Cornuet et al., 1978).

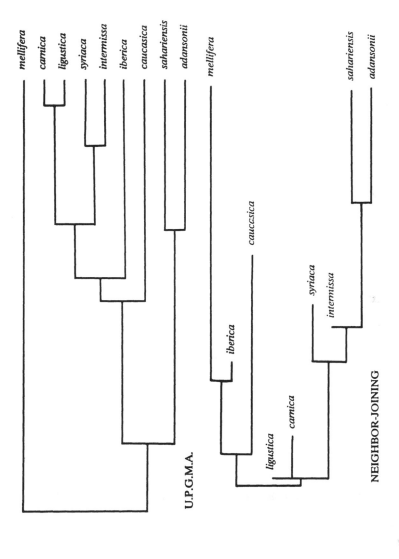

Figure 6.2 Phenograms for nine *Apis mellifera* subspecies generated by UPGMA and the Neighbor-joining method of Saitou and Nei (1987) using as data the Mahalanobis D2 distances between the centroids from the factorial discriminant analysis presented in Figure 6.1.

With morphometry, *Apis mellifera* appears as a highly differentiated species. Most of the 24 races or subspecies described by Ruttner (1988) can be easily discriminated by biometric data. However, the distinction between some African subspecies may be quite difficult. *A. m. capensis* and *scutellata* are two distinct subspecies with different behavior and reproductive biology, but their external morphological characters are very similar. Also, the differentiation of populations seems to be much less among African subspecies than among certain European subspecies (Gadbin et al., 1979).

Because of the undefined relationship between phenotypes and genotypes, morphometric characters are not good indicators of phylogenetic links between these taxa. However Ruttner (1988) deduced a kind of phylogenetic tree of the 24 subspecies from the result of a PCA analysis. In his three-dimensional phenogram he postulates a four-branched radiation (African, west-Mediterranean, north-Mediterranean, Caucasian). Using the Mahalanobis D2 distances between the centroids of the 9 subspecies of Figure 6.1, we can build "phylogenetic" trees according to different algorithms (UPGMA and the Neighbor-joining of Saitou and Nei, 1987). Figure 6.2 presents both trees. They differ mainly in the branching of *A. m. iberica* which appears much more distant from *A. m. mellifera* in the Neighbor-joining tree than in the UPGMA tree.

Allozyme Variation

Since the first discovery of a biochemical polymorphism in honey bees by Mestriner some 20 years ago, many authors have published data on allozyme variation in this species. Even if new polymorphic loci are found from time to time (so far, six enzymes have shown some electrophoretic variants), the overall level of allozyme variation is low. Four of these six loci, aconitase (Acon), hexokinase (Hk), malic enzyme (Me), and phosphoglucomutase (Pgm), present either localized or rare variants. The esterase (Est) locus presents more variation: 3 alleles are commonly found in *A. m. scutellata* (Sheppard and Heuttel, 1988) and in the European subspecies *A. m. carnica* (Sheppard and McPheron, 1986), *A. m. ligustica* (Badino et al., 1983) and *A. m. cecropia* (Badino et al., 1988). Up to four alleles have been found in *A. m. sicula* (Cornuet, unpublished results). The most informative locus is unquestionably the cytoplasmic malate dehydrogenase locus (Mdh1), for which at least 6 alleles have been found: three common ones, $Mdh1^{65}$, $Mdh1^{80}$ and $Mdh1^{100}$ (Sheppard and Berlocher, 1984, 1985) and three rarer ones, called $Mdh1^{55}$ (Sheppard and McPheron,

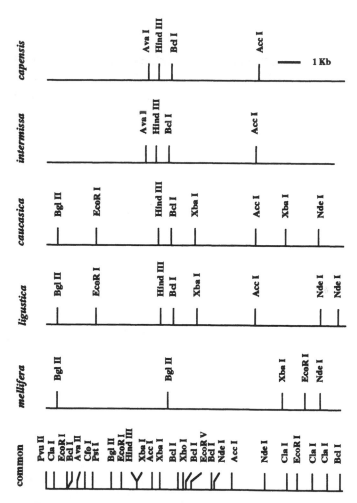

Figure 6.3 Restriction enzyme cleavage site maps for the mitochondrial DNA of *Apis mellifera mellifera, ligustica, caucasica, intermissa* and *capensis*. Restriction sites common to all subspecies are shown at the bottom.

1986), *Mdh1*[87] (Sheppard and Berlocher, 1985) and an allele that migrates much ahead of *Mdh1*[100] called *F1* (Badino et al., 1985), which we also found in colonies of *A. m. ligustica*. The rare alleles are each generally found in discrete localities: e.g., *Mdh1 F1* in eastern Sicily and Calabria, Italy (Badino et al. 1985), *Mdh1*[87] in Forli, Italy (Sheppard and Berlocher, 1985) and *Mdh1*[65] in Norway (Sheppard and Berlocher, 1984). In some colonies from the south of France we also found a very slow allele, that is perhaps *Mdh1*[55].

The three main Mdh1 alleles give a coherent picture of the species compatible with the Ruttner's hypothesis, outlined above. The Caucasian and African branches have only (or mainly) the $Mdh1^{100}$ allele. This allele is progressively replaced by the $Mdh1^{80}$ allele along the west-Mediterranean branch. At the end of this branch the $Mdh1^{100}$ allele has virtually disappeared; it is found only in places where genetic pollution by other subspecies cannot be ruled out. The $Mdh1^{65}$ allele plays an identical role in the north-Mediterranean branch, but the $Mdh1^{100}$ allele is still present at a frequency of around 0.30.

The tree obtained by Sheppard and Heuttel (1988) using data from four loci, Mdh1, Est, Acon, and Pgm, and the five subspecies *A. m. mellifera, carnica, ligustica, capensis* and *scutellata* is in good agreement with those obtained from morphometric data.

DNA Variation

In the last three years the first publications on variation at the DNA level in honey bees appeared. The general technique is to look for restriction fragment length polymorphisms (RFLPs). It has been applied to nuclear DNA (Hall 1986, 1988; Severson et al., 1988) as well as to mitochondrial DNA (Moritz et al., 1986; Smith, 1988, 1991; Smith and Brown, 1988, 1990; Smith et al., 1989, 1991). Mitochondrial DNA (mtDNA) has recently been used as a marker to distinguish between male and female influence in the Africanization process in the Americas (Smith et al., 1989; Hall and Muralidharan, 1989; Hall and Smith, 1991).

Besides its potential discriminating power, mtDNA may provide valuable information on the phylogenetic links between subspecies or populations. So far the restriction maps of a few subspecies have been published. Here we will give the position of sites for restriction enzymes that have not yet been published and the map of an additional subspecies, *A. m. intermissa*.

Methods and Materials

The following *A. mellifera* subspecies were sampled: 8 colonies of *A. m. mellifera* from three locations in continental France and 1 location in Corsica; four colonies of *A. m. caucasica*; four colonies of *A. m. ligustica*, two colonies of *A. m. intermissa* from Morocco and two from Algeria; and one colony of *A. m. capensis*.

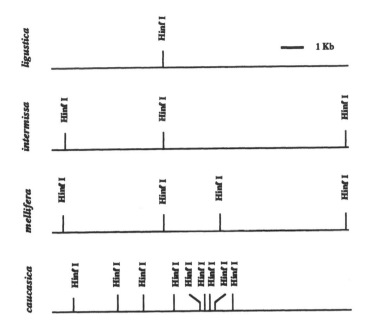

Figure 6.4 Maps of *Hinf* I cleavage sites in the mitochondrial DNAs of *Apis mellifera caucasica, mellifera, intermissa* and *ligustica*.

Newly emerged bees are ground in ice-cold TE buffer with 15% sucrose. After two rounds of centrifugation (5 minutes at 1000 x *g* and 5 minutes at 2500 x *g*), mitochondria are pelleted at 10,000 x *g* during 20 min. Then they are purified on a discontinuous sucrose gradient (20%-30%-42.5%). The mitochondrial membranes are lysed in 1% SDS for 5 min at room temperature. Proteins are removed by two phenol-chloroform extractions and nucleic acids are precipitated overnight in sodium acetate/ethanol at -20° C. After pelleting and rinsing in 75% ethanol, they are resuspended in TE buffer or sterile water.

With this protocol the greatest yield of mtDNA is obtained with newly emerged bees. With older bees, the yield is drastically reduced. The following restriction enzymes have been assayed in all colonies: *Acc* I, *Ava* I, *Ava* II, *Bcl* I, *Bgl* II, *Cla* I, *Cfo* I, *Eco*R I, *Eco*R V, *Hin*d III, *Hinf* I, *Nde* I, *Pst* I, *Pvu* II, and *Xba* I. The restriction enzymes *Bam*H I and *Sca* I have been tested in one colony for each of the three subspecies *mellifera*, *ligustica* and *caucasica*. As no restriction sites were found, these enzymes were not subsequently used.

For most of the digested sample, electrophoresis was performed on agarose (0.8%) and polyacrylamide (3.6%) gels. DNA bands were detected either by fluorescence in ethidium bromide, by silver nitrate

Table 6.1 Nucleotide divergence among the mitochondrial genomes of *Apis mellifera* subspecies. Pairwise comparisons with *A. m. capensis* have been calculated without *Hinf* I sites. See text for discussion.

	mellifera	ligustica	caucasica	intermissa
ligustica	2.52			
caucasica	2.3	0.65		
intermissa	2.9	1.65	1.34	
capensis	[2.62]	[1.83]	[2.10]	[0.00]

staining (Tegelström, 1986) or by autoradiography of end-labeled DNA.

Double digests were used to establish restriction maps. The sequence published by Crozier et al. (1989) helped us to map the numerous *Hinf* I sites. The nucleotide divergence between mtDNAs in pairwise populations has been computed according to the method of Nei and Tajima (1983).

Results

Pooling all samples, a total of 52 sites have been mapped, 13 of which are *Hinf* I sites. Probably because of very small sample size we did not detect any restriction site variation within subspecies. However, as previously found by others, we did find restriction site variation among subspecies. Figure 6.3 summarizes our data for the non-*Hinf* I sites. Differences have been found between all pairs of subspecies with one exception: the two African subspecies *intermissa* and *capensis* have 30 common sites. Figure 6.4 shows the restriction map for *Hinf* I. All the sites found in *A. m. caucasica* are found in the other subspecies, but all subspecies differ by additional site(s). We have not yet mapped *Hinf* I sites in *A. m. capensis*.

Table 6.1 gives the nucleotide divergence among the five subspecies studied. The divergence data for *A. m. capensis* are not comparable with those of other subspecies since the sample of restiction sites is reduced. In fact, the difference between the last two lines of the table is entirely due to effects of the *Hinf* I sifes. Figure 6.5 presents the phylogenetic trees deduced from nucleotide divergence matrix. If *capensis* and *intermissa* have similar *Hinf* I sites they would branch very high on the tree.

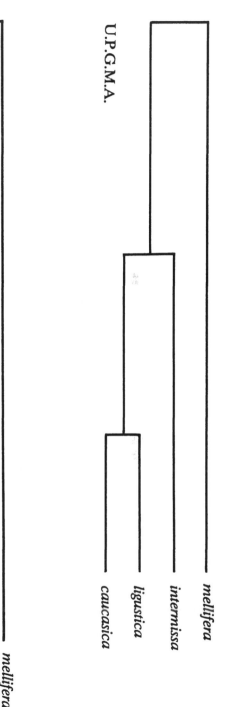

U.P.G.M.A.

NEIGHBOR-JOINING

Figure 6.5 Phenograms for four *Apis mellifera* subspecies generated by the UPGMA and Neighbor-joining methods, using as data percent sequence divergence among mitochondrial genomes.

Discussion

Restriction maps of *A. m. mellifera, ligustica, caucasica, scutellata* and *capensis* have already been published by D. Smith and co-workers (Smith, 1988; Smith and Brown, 1988, 1990; Smith et al., 1991). Generally they are in good agreement with ours. The few differences can be explained by variation within subspecies. For instance, Smith et al. (1989) show two *Xba* I sites for *A. m. ligustica* which we did not find; however these two sites are noted as being polymorphic within *A. m. ligustica*.

We mapped several new restriction sites, generated by the enzymes *Ava* II, *Cfo* I, *Cla* I and *Hinf* I. The first three revealed 6 sites that appeared to be constant in all tested samples. Among the four-base restriction enzymes we tried *Alu* I, *Dde* I, *Mbo* I and *Hinf* I. All produce variable profiles between and also within subspecies, but only *Hinf* I, which produced the fewest restriction fragments, could be mapped. This enzyme, taken alone, gives discrimination among the principal subspecies.

General Discussion

Apis mellifera is a highly differentiated species. This is especially revealed by morphometric techniques which, in spite of their imprecision at the genetic level, are generally the most powerful tool for discrimination among populations. Even if new enzymatic polymorphisms are found, morphometrics will remain more useful than allozymes in separating subspecies or populations. Some exceptions exist. In France, for instance, the MDH1 locus is used to measure the genetic pollution of the local subspecies, *A. m. mellifera* (Cornuet et al., 1986). In its natural state, this subspecies has only one MDH1 allele and imported queens are typically from populations (such as Italy, Austria and Russia) where this allele does not exist. When the level of genetic pollution is low, morphometrics is not suited for its detection.

The level of diversity in mtDNA is higher than for allozyme variation. Most of the subspecies studied so far can be differentiated. However, no diagnostic enzyme has yet been found to distinguish between *ligustica* and *carnica*, or, more surprisingly, between *intermissa* and *capensis*. Maybe a four-base restriction enzyme such as *Hinf* I will bring the answer. With its matrilineal transmission, mtDNA is not sensitive to genetic pollution. This is an advantage when one wants to determine the original mitotype of a population. But, when considering one particular colony alone, the mitotype alone does not allow determination of its genetic origin.

of a population. But, when considering one particular colony alone, the mitotype alone does not allow determination of its genetic origin.

Morphometric data are not well suited for inferring phylogenies since powerful selective factors may result in convergence. For instance, the body size and length of hairs vary gradually along climatic gradients (Ruttner, 1988). The trees presented in Figure 6.2 are only for comparative purpose. However, Ruttner based his evolutionary hypothesis not only on paleoclimatic and paleogeographic knowledge but also on morphometrical data.

MtDNA is much better suited for this kind of study. The overall picture is conserved but two unexpected results have not received a definitive interpretation.

First, the level of nucleotide divergence is much higher between some subspecies than expected under Ruttner's phylogenetic hypothesis. According to Ruttner (1988) *Apis mellifera* is a young species that separated from *Apis cerana* only 50,000 years ago. The differentiation in races would have occured during the last glaciation, some 10,000 years ago. An order of magnitude for the speed of nucleotide divergence between isolated populations is about 1%-2% per million years. Even if we take into account the fact that mtDNA divergence need not be synchronous with population separation, there still exists a discrepancy: there is too much sequence divergence for the proposed age of the taxa. We propose three, non-mutually exclusive explanations:

1. Our estimations of sequence divergence are biased, due to the choice of restriction enzymes. However, our estimates are slightly lower than those of Smith (e.g., Smith, 1990), probably because we have included additional restriction enzymes which produce less variable profiles in average (6 constant sites for *Cla* I, *Ava* II and *Cfo* I).
2. MtDNA has evolved very rapidly in honey bees. This is supported by sequence analysis of the COI-COII region of honey bee mtDNA (Crozier et al., 1989) which shows that this sequence has evolved about twice as fast in *Apis* as in *Drosophila*.
3. The separation between *A. m. mellifera*, the African subspecies, and *A. m. ligustica* occurred in more ancient times than previously thought. This is paralleled by the result of a recent allozyme study (Sheppard and Berlocher, 1989) suggesting that the *Apis mellifera/ Apis cerana* cladogenesis did not take place in the Quaternary (approximately 2 million years ago) as has often been suggested, but much earlier, perhaps in the mid-Tertiary era.

A second surprising result is the high similarity between north- and south-African subspecies. We have not yet any satisfying explanation for that phenomenon. Nevertheless, this finding is hardly compatible with Ruttner's hypothesis of close relationships between *intermissa, iberica* and *mellifera*, which together comprise the west-Mediterranean branch recognized by morphometric studies.

Acknowledgments

We wish to thank R. F. A. Moritz for providing our *A. m. capensis* sample. We are very grateful to M. Solignac and M. Pélandakis for the multiple and very helpful discussions on many mtDNA aspects.

Literature Cited

Badino, G., G. Celebrano and A. Manino. 1983. Population structure and *Mdh-1* locus variation in *Apis mellifera ligustica. J. Hered.* 74:443-446.

Badino, G., G. Celebrano, A. Manino and M. D. Ifantidis. 1988. Allozyme variability in Greek honeybees (*Apis mellifera* L.). *Apidologie* 19:377-386.

Badino, G., G. Celebrano, A. Manino and S. Longo. 1985. Enzyme polymorphism in the Sicilian honeybee. *Experientia* 41:752-754.

Cornuet, J.-M., A. Daoudi and C. Chevalet. 1986. Genetic pollution and number of matings in a black honey bee (*Apis mellifera mellifera*). *Theor. Appl. Genet.* 73:223-227.

Cornuet, J.-M., J. Fresnaye and P. Lavie. 1978. Etude biométrique de deux populations d'abeilles cévenoles. *Apidologie* 9:41-55.

Crozier, R. H., Y. C. Crozier and A. G. McKinley. 1989. The CO-I and CO-II region of honeybee mitochondrial DNA: evidence for variation in insect mitochondrial evolutionary rates. *Mol. Biol. Evol.* 4:399-411.

Gadbin, C., J.-M. Cornuet and J. Fresnaye. 1979. Approche biométrique de la variété locale d' *Apis mellifica* L. dans le sud tchadien. *Apidologie* 10:137-148.

Hall, H. G. 1986. DNA differences found between Africanized and European honeybees. *Proc. Natl. Acad. Sci. USA* 83:4874-4877.

------. 1988. Characterization of the African honey-bee genotype by DNA restriction fragments. In *Africanized Honey Bees and Bee Mites*, G. R. Needham, R. E. Page, Jr., M. Delfinado-Baker and C. E. Bowman, eds. Ellis Horwood: Chichester. Pp.287-293.

Hall, H. G. and K. Muralidharan. 1989. Evidence from mitochondrial DNA that African honey bees spread as continuous maternal lineages. *Nature* 339:211-213.

Hall, H. G. and D. R. Smith. 1991. Distinguishing African and European honey bee matrilines using amplified mitochondrial DNA. *Proc. Natl. Acad. Sci. USA* in press.

Moritz, R. F. A., C. F. Hawkins, R. H. Crozier and A. G. McKinley. 1986. A mitochondrial DNA polymorphism in honeybees (*Apis mellifera* L.) *Experientia* 42:322-324.

Nei, M. and F. Tajima. 1983. Maximum likelihood estimation of the number of nucleotide substitutions from restriction sites data. *Genetics* 105:207-217.

Ruttner, F. 1988. *Biogeography and Taxonomy of Honeybees.* Springer-Verlag: Berlin, Heidelberg.

Ruttner, F., L. Tassencourt and J. Louveaux. 1978. Biometrical statistical analysis of the geographic variability of *Apis mellifera* L. I: Material and methods. *Apidologie* 9:362-382.

Saitou, N. and M. Nei. 1987. The neighbor-joining method: a new method for reconstructing phylogenetic trees. *Mol. Biol. Evol.* 4:406-425.

Severson, D. W., J. M. Aiken and R. F. Marsh. 1988. Molecular analysis of North American and Africanized honey bees. In *Africanized Honey Bees and Bee Mites*, G. R. Needham, R. E. Page, Jr., M. Delfinado-Baker and C. E. Bowman, eds. Ellis Horwood: Chichester. Pp. 294-302.

Sheppard, W. S. and S. H. Berlocher. 1984. Enzyme polymorphism in *Apis mellifera* from Norway. *J. Apic. Res.* 23:64-69.

------. 1985. New allozyme variability in Italian honey bees. *J. Hered.* 76:45-48.

------. 1989. Allozyme variation and differentiation among four *Apis* species. *Apidology* 20:419-431.

Sheppard, W. S. and M. D. Huettel. 1988. Biochemical genetic markers, intraspecific variation, and populations genetics of the honey bee, *Apis mellifera.* In *Africanized Honey Bees and Bee Mites*, G. R. Needham, R. E. Page, Jr., M. Delfinado-Baker and C. E. Bowman, eds. Ellis Horwood: Chichester. Pp. 281-286.

Sheppard, W. S. and B. A. McPheron. 1986. Genetic variation in honey bees from an area of racial hybridization in western Czechoslovakia. *Apidology* 17:21-32.

Smith, D. R. 1988. Mitochondrial DNA polymorphisms in five Old World subspecies of honey bees and in New World hybrids. In *Africanized Honey Bees and Bee Mites*, G. R. Needham, R. E. Page, Jr., M. Delfinado-Baker and C. E. Bowman, eds. Ellis Horwood: Chichester. Pp. 303-312.

------. 1991. African bees in the Americas: insights from biogeography and genetics. *TREE* 6:17-21.

Smith, D. R. and W. M. Brown. 1988. Polymorphisms in mitochondrial DNA of European and Africanized honeybees (*Apis mellifera*). *Experientia* 44:257-260.

------. 1990. Restriction endonuclease cleavage site and length polymorphisms in mitochondrial DNA of *Apis mellifera mellifera* and *A. m. carnica* (Hymenoptera: Apidae). *Ann. Entomol. Soc. Am.* 83:81-88.

Smith, D. R., Palopoli, M. F., Taylor, B. R., Garnery, L., Cornuet, J.-M., Solignac, M. and Brown, W. M. 1991. Geographic overlap of two classes of mitochondrial DNA in Spanish honey bees (*Apis mellifera iberica*). *J. Hered.* 82:96-100.

Smith, D. R., O. R. Taylor and W. M. Brown. 1989. Neotropical Africanized honeybees have African mitochondrial DNA. *Nature* 339:213-215.

Tegelström, H. 1986. Mitochondrial DNA in natural populations: An improved routine for screening of genetic variation based on sensitive silver staining. *Electrophoresis* 7:226-229.

7

Allozyme Diversity in Asian *Apis*

Gan Yik-Yuen, Gard W. Otis, Makhdzir Mardan,
and Tan S. G.

Introduction

Gel electrophoresis of enzymes, the protein products of genes, provided biologists with the first technique that allowed them to examine genetic diversity directly. It was soon recognized that this unanticipated wealth of genetic variability at both the population and species levels provided new sets of characters that could be useful for phylogenetic inference (Avise, 1974; Buth, 1984). Thus, the discovery of numerous electrophoretic allozymes opened up a new field of molecular taxonomy. Recent techniques of DNA analysis that allow more direct examination of the genome have largely eclipsed the use of allozyme data for systematics. It is generally accepted, however, that allozymes have been and still are of value in systematics (Avise, 1974; Buth, 1984; Hillis, 1987, and references therein), particularly in the detection of cryptic species (Avise, 1974; Berlocher, 1984; Eldridge et al., 1986). However, it is also clear that, as with morphological characters, there can be convergences in molecular data sets as well as variation that lacks phylogenetic significance (Hillis, 1987).

The western honey bee, *Apis mellifera* L., has been the subject of a great number of allozyme studies. While many of these are descriptive studies of the bees from different geographic regions, others have demonstrated the possibility of using allozymes for racial discrimination (Nunamaker et al., 1984b; Sylvester, 1982, 1986; Daly, 1991 and references therein), the existance of hybrid zones between two races of bees (Sheppard and McPheron, 1986), and the probable operation of the founder effect (Cornuet, 1979; Sheppard, 1988). More recently allozymes have been used as genetic markers for making blind observations of bee behavior (Robinson and Page, 1988; Robinson and Page, 1989; Robinson et al., 1990). Sheppard and Huettel (1988) constructed a phylogeny of *A. mellifera* based on genetic distance matrices generated from allozyme data.

These studies generally agree that *Apis mellifera* exhibits relatively little allozyme variation. Mean heterozygosity has been estimated as only 0.010 - 0.012 (Metcalf et al. 1975; Pamilo et al., 1978). To date only six enzymes have been reported to be polymorphic despite extensive surveying. This has limited the usefulness of allozymes for most of these studies.

Surprisingly, the genus *Apis* as a whole has been generally neglected with respect to allozyme studies. With *A. mellifera* being a relatively recent species (1-2 million years, Ruttner, 1988), it is perhaps not surprising that it does not show extensive allozyme variation. One might expect to find more variation in more ancient species or in populations occupying the center of the geographic range of the genus. In addition, until recently only four species were recognized within the genus: a dwarf honey bee, a giant honey bee, and two cavity-nesting species. These four taxa have been extensively reviewed by Ruttner (1988). The recent recognition of at least one additional species in each of these three groups (see Chapter 2, this volume) has greatly increased the recognized diversity within *Apis*, which in turn has been responsible in part for the renewed interest in the phylogenetic relationships within the honey bees (Chapter 1, this volume).

In this chapter, we first review the published studies on allozymes of Asian honey bees. The main emphasis of the chapter will be a summary of our recent studies on allozymes of bees from Thailand, Peninsular Malaysia, Borneo, the Philippines, and Sulawesi. We will conclude with a reassesment of the genus and some directions for future research.

Previous Studies of the Allozymes of Asian Honey Bees

To date there have been only four studies comparing the allozymes of Asian honey bees, each focusing on bees from the edges of the species' ranges. All are recent, with the first by Nunamaker, Wilson, and Ahmad (1984a) being conducted on specimens of *Apis cerana, A. dorsata*, and *A.*

florea collected in Pakistan. They individually analyzed 100 individuals from 10-12 colonies of each species, by means of isoelectric focusing in thin layer polyacrylamide gels. Two enzymes were studied: MDH and EST. (The abbreviation for the enzyme produced by a gene locus is given in capital letters: MDH; the gene locus is in capital and lower-case letters: Mdh; alleles of the locus are in italic capitals and lower-case: Mdh^{100}. See Table 7.1 for enzyme names and abbreviations). MDH produced one intense band in *florea* and two bands, one intense and one faint, in both *dorsata* and *cerana*. In contrast, for non-specific esterases, *florea* and *dorsata* shared two intense bands, but had different faint bands (three in *florea*, one in *dorsata*). A. *cerana* had two faint bands which were not shared with either of the other two species. The study detected no intraspecific polymorphisms. This lack of variation may have resulted from the relatively small number of colonies sampled from populations at the edge of the species' range.

A brief report on EST enzymes by Tanabe and Tamaki (1985) also indicated that *mellifera* and *cerana* had species specific differences. In the zymograms obtained through agarose gel electrophoresis, both species had four bands, but in each species a different one was densest. These banding patterns differ from those reported by Nunamaker et al. (1984a).

Li et al. (1986) studied EST enzymes of honey bees and several other genera of bees in southern Yunnan Province, China, using isoelectric focusing on polyacrylamide gels. Approximately 100 specimens of *cerana* and *mellifera* were studied; in contrast, only about 20 individuals of the other four species were analyzed (*florea, dorsata, andreniformis,* and *laboriosa*). These latter two species, and perhaps the others as well, were represented by bees from a single colony. Each species had a unique zymogram, although matching of the bands between species is difficult because representatives of the different species do not appear to have been run simultaneously on the same gel. As in the previous study, there was no intraspecific variation, perhaps due to the relatively small number of colonies that were sampled. The collection sites were also on the edge of the ranges of all but the cavity-nesting species, A. *cerana*.

Sheppard and Berlocher (1989) conducted a more extensive survey of the allozymes of honey bees from Sri Lanka. They analyzed fifteen bees from each of ten colonies of *cerana* and *florea*, and from each of five colonies of *dorsata*, from sites throughout the island country. Horizontal starch gel electrophoresis was used to survey bees for seventeen enzymes, which represented eighteen putative loci; details are given in their paper. Three loci, Lap, Pgi and β-Hbdh, were fixed for one allele across all species. Another eight loci -- Acon-1, Aldo, α-Gdh, Ao, Argk, Idh, Idhn,

Table 7.1 Enzymes mentioned in text, with abbreviations and Enzyme Commision numbers.

ENZYME	ABBREVIATION	E.C. #
aconitase	ACON-1, ACON-2	4.2.1.3
acid phosphatase	ACP	3.1.3.2
aldehyde dehydrogenase	ALDH	1.2.1.3
aldolase	ALDO	4.1.2.13
alkaline phosphatase	ALP	3.1.3.1
aldehyde oxidase	AO	1.2.3.1
arginine kinase	ARGK	3.3.8.9
esterases	EST	3.1.1.1
fumarase	FUM	4.2.1.2
glyceraldehyde-3-phosphate dehydrogenase	G-3-PDH	1.2.1.12
glyoxalase-1	GLO-1	4.4.1.5
glucose dehydrogenase	GLDH	1.1.1.47
α-glycerophosphate dehydrogenase	α-GDH	1.1.1.8
β-hydroxybutyrate dehydrogenase	β-HBDH	1.1.1.30
hexokinase	HEX	2.7.1.1
isocitrate dehydrogenase	IDH	1.1.1.42
NAD-dependent isocitrate dehydrogenase	IDHN	1.1.1.41
leucine amino peptidase	LAP	3.4.1.1
malate dehydrogenase	MDH	1.1.1.37
malic enzyme	ME	1.1.1.40
6-phosphogluconate dehydrogenase	6-PGD	1.1.1.43,44
phosphoglucose isomerase	PGI	5.3.1.9
phosphoglucomutase	PGM	2.7.5.1
shikimate dehydrogenase	SHDH	1.1.1.25
succinate dehydrogenase	SUDH	1.3.99.1
triose phosphate isomerase	TPI	5.3.1.1

and G-3-Pdh -- showed variation between some of the species, but were not polymorphic within a species. They found seven polymorphic loci in the Asian honey bees. The loci and the number of putative alleles were: *A. cerana*: Acon-2 (2), Est (2), Mdh (2), and Me (2); *A. dorsata*: Hex (2); *A. florea*: Me (2), Pgm (3), and Tpi (2). These same enzymes, with the exception of TPI, have been found to be polymorphic in *A. mellifera* as well. In most of these enzymes, one allozyme (putative allele) was common for a particular species while the alternative allele(s) were rare. In some cases, Sheppard and Berlocher recorded a locus as polymorphic even though the rare allele was found in only one bee out of 150 analyzed for the species (e.g., *cerana*, Est); they seem to have adopted a less stringent definition of "polymorphic" than used by many authors (e.g.,

Hartl, 1988). Only in the case of Me and Tpi in *A. florea* were the alternative alleles relatively common. Consequently, the mean heterozygosity for each species was very low.

Sheppard and Berlocher (1989) calculated a mean Nei's genetic distance (D) of 1.30 between the four traditional species of *Apis*. This indicates that these species are distinct, with substantial differentiation between each. Even between *mellifera* and *cerana*, the differences are extensive (D = 1.10), and they suggest that the divergence of these two species may predate the generally accepted time of 1-2 million years ago. Because of the large number of alleles showing fixed differences between species, the allozyme data were not useful in resolving questions concerning the phylogeny of the honey bees. Their UPGMA analysis supported an early divergence time for *A. florea*, but was also of no help in resolving the relationships between the other three species. Because each species had a different allozyme profile, Sheppard and Berlocher argued that electrophoretic techniques should be useful in differentiating potential species of honey bees (see Chapter 2, this volume).

The Malaysian Study of
Allozymes of Asian Honey Bees

In 1988, Makhdzir Mardan of the Universiti Pertanian Malaysia (UPM) provided specimens of *Apis cerana*, *A. dorsata*, and *A. koschevnikovi* to Gan Yik Yuen for preliminary allozyme analyses. The results indicated there was substantial polymorphism which seemed promising to quantify in more detail. While based at UPM during the first half of 1989, Gard Otis made extensive collections of bees in several locations in Southeast Asia, which he made available to Gan and other researchers. She, in turn, devoted a major portion of her energies and financial resources to electrophoresing the relatively large number of samples of bees. Mardan and Tan provided logistic support throughout the project. The following is a report on the results of this collaborative work.

When we initiated this study, none of us had any previous experience working with the enzymes of honey bees. This probably was to our advantage because we had no biases concerning which enzymes we expected to be polymorphic, other than MDH. The set of enzymes that we examined was independently decided upon by Gan, based on extensive experience with electrophoresis of other organisms.

Adult worker bees were collected from Bangkok (*cerana, dorsata, florea*); Peninsular Malaysia (*cerana, dorsata, andreniformis*); Sabah,

Borneo (*cerana, dorsata, andreniformis, koschevnikovi*); Luzon, Philippines (*mellifera* imported from Australia, *cerana*); South Sulawesi, Indonesia (*cerana, binghami*); and Guelph, Ontario (*mellifera*). The bees were stored soon after collection in liquid nitrogen, then stored in a freezer (-40°C) at UPM until the time of analysis.

A preliminary survey was performed to compare the enzyme profiles obtained from different body parts of the bees and check for tissue-specific expression. The results indicated that the enzyme patterns obtained from the head, thorax, abdomen, and legs were identical. Hence, a section of these body parts was used each time to type for various enzymes.

The honey bee tissue was homogenized in approximately equal volume of ice-cold distilled water. Electrophoresis was immediately conducted on this crude extract using 7% polyacrylamide horizontal slab gels. Forty samples arranged in two tiers were inserted into each gel. Altogether 22 different gel and electrode buffers were systematically tried out for each enzyme. The best buffer systems for each enzyme were selected (Table 7.2). The staining procedures used were from Shaw and Prasad (1970), Harris and Hopkinson (1976), Munstermann (1979), Steiner and Joslyn (1979), Menken (1980, 1982) and Varvio-Aho and Pamilo (1980).

The summary that follows presents the general results of our research. In this preliminary analysis of the electrophoretic data, we detected a large number of enzymes which were either polymorphic within a species, or which exhibited fixed differences between species. For all but esterase we present the data using letters to denote allozymes (alleles). Only Me, Alp, Acp, and 6-Pgd were completely monomorphic. (After reading Sheppard and Berlocher's (1989) report of polymorphism in malic enzyme, we also found this enzyme to be polymorphic when using a different buffer, but we had no opportunity to completely reassess Me in this study.) The status of each isozyme (e.g., polymorphic, monomorphic) is presented in the last column of Table 7.2. More detailed reports on each of the polymorphic enzymes, including relative mobilities, will be the subjects of future research papers.

Esterase (EST)

Our analyses suggest the presence of two separate esterase isozymes, probably the products of two loci, Est-1 and Est-2. Est-1 expression was found primarily in *mellifera* and *cerana*; it was either very weak or not expressed in the other species. *A. mellifera* had a high frequency of the Est-1^{100} allele and a low proportion of Est-1^{70}; the EST-2 isozyme was not observed. In *cerana*, Est-1^{70} was the most common allele; a few bees had Est-1^{100}. The EST-2 isozyme was recorded in only a few *cerana* bees

Table 7.2 Summary of enzymes and buffer systems surveyed by Gan Y. Y. and associates.

Enzyme	Buffer System	Reference	Status
EST	CA-7	Steiner & Joslyn, 1979	polymorphic
MDH	TEMM	Spencer et al., 1964	polymorphic
GLDH	Menken-4	Menken, 1980, 1982	polymorphic
α-GDH	TEMM		polymorphic*
FUM	V-Aho	Varvio-Aho & Pamilo, 1980	polymorphic*
SUDH	TEMM		polymorphic*
SHDH	TEMM		polymorphic*
ME	TC-3	Gan, unpublished[2]	monomorphic[1]
ALP	TC-7	Gan, unpublished[2]	monomorphic
ACP	TC-7		monomorphic
6-PGD	TEMM		monomorphic
IDH	CA-7		polymorphic*
TPI	V-Aho		polymorphic
ALDH	CA-7		polymorphic
GLO-1	CA-7	Steiner & Joslyn, 1979	polymorphic

* newly reported polymorphism;
[1] polymorphic using a different buffer (see text for discussion)
[2] Tris-citrate buffer; available from Gan Y. Y.

from Thailand, Sabah, and Sulawesi.

The open-nesting species and *koschevnikovi* all showed strong expression of Est-2 (*koschevnikovi*: $Est\text{-}2^{180}$; *florea*: $Est\text{-}2^{150}$; *andreniformis*: $Est\text{-}2^{170}$; *dorsata* and *binghami*: $Est\text{-}2^{160}$). Between 22-76% of the bees in these species could not be scored for EST-1. When it was present, these species had predominantly the $Est\text{-}1^{70}$ allele, although all but *florea* also had a rare $Est\text{-}1^{80}$ band not recorded in *mellifera* or *cerana*. The Est-2 allozymes were only 1.5 mm apart on the gels. Consequently, many samples were electrophoresed a second time. The initial results were repeatable and therefore the validity of these allozymes was confirmed.

Malate Dehydrogenase (MDH)

This is one of the most polymorphic isozymes in *Apis mellifera*. Not surprisingly, we recorded six different alleles. The A and B allozymes were specific to the cavity-nesting species (*mellifera, cerana, koschevnikovi*). *Apis koschevnikovi* had only the B allele. The dwarf bees were both monomorphic but for different alleles (*florea*, C; *andreniformis*, D). *Apis dorsata* and *binghami* shared a common allele (F) and a rare allele (E), and each had one additional rare allele (C and D respectively in the two species).

Glucose Dehydrogenase (GLDH)

We recorded three allozymes shared by all three species at this locus. The A and B allozymes were common in *dorsata* and *binghami*. In all other species B and C alleles were the more common.

α-Glycerophosphate Dehydrogenase (α-GDH)

This isozyme was found to be polymorphic in *Apis* for the first time. Three different allozymes were found, but one was rare and found only in *cerana* and *koschevnikovi*. For the other two allozymes (A and B), *florea* and *andreniformis* were monomorphic for A while *mellifera* was monomorphic and *cerana* and *koschevnikovi* were nearly so for B. The two species of giant bees shared the A and B alleles with high frequencies of each.

Fumarase (FUM)

This isozyme has not previously been reported to be polymorphic. Surprisingly, we found it to have a great deal of variability with five distinct alleles. All species had the A, B, and C alleles in differing proportions. For *mellifera* we recorded frequencies of 0.30, 0.43, and 0.24 respectively. In contrast, the combined frequencies of these three allozymes in *cerana* was only 0.11. *A. florea* and *koschevnikovi* had predominantly the B allozyme, while *andreniformis* had a high frequency of the C allozyme. The E allele was the predominant one in *cerana*, *dorsata*, and *binghami*, and was recorded at low frequency in *andreniformis*. The D allele was generally rare and was recorded in *mellifera*, *cerana*, *dorsata*, and *binghami* (maximum frequency = 0.06). For unexplained reasons there was a paucity of heterozygotes recorded for Fum.

Succinate Dehydrogenase (SUDH)

This new polymorphic isozyme had five allozymes. The most variable species was *mellifera* with four allozymes having frequencies greater than 0.05 (A, B, C, and D); the A allozyme predominated. For each of the other species, there was one predominant allozyme (B for *koschevnikovi* and *florea*; C for *andreniformis*; and E for *cerana*, *dorsata*, and *binghami*) and one or two rarer alleles (maximum frequency = 0.10).

Table 7.3 Comparison of the observed heterozygosities in seven species and subspecies of *Apis*. M = *A. mellifera*; C = *A. cerana*; K = *A. koschevnikovi*; F = *A. florea*; A = *A. andreniformis*; D = *A. dorsata dorsata*; B = *A. dorsata binghami*. See Table 7.1 for enzyme abbreviations and Enzyme Commission numbers.

Enzyme	M	C	K	F	A	D	B
EST-1	0.107	0.031	*	*	*	*	*
EST-2	*	*	0.024	0.038	0.030	0.035	0.056
MDH	0.760	0.123	0	0	0	0.087	0.158
GLDH	0.524	0.605	0.479	0.386	0.486	0.433	0.300
α-GDH	0	0.067	0.061	0	0	0.823	0.500
FUM	0.396	0.179	0.179	0.160	0.140	0.481	0.687
SUDH	0.44	0.046	0	0.033	0	0.029	0.200
SHDH	0	0	0	0	0.212	0.071	0
ALP	0	0	0	0	0	0	0
ACP	0	0	0	0	0	0	0
6-PGD	0	0	0	0	0	0	0
AVERAGE	0.233	0.105	0.074	0.062	0.087	0.196	0.211

* For any particular species, either EST-1 or EST-2 could not be consistently scored; therefor no heterozygosity value is given.

Shikimate Dehydrogenase (SHDH)

This isozyme has never been reported as polymorphic for honey bees. In fact, it is rarely encountered in organisms other than plants. We found the same two allozymes in all species, with one consistently being found at higher frequencies than the other.

Other Enzymes

Me, Alp, Acp and 6-Pgd were all monomorphic, although as indicated malic enzyme was found to be polymorphic later with different buffer systems. Idh was found to be polymorphic for the first time; other enzymes that were successfully scored were Aldh, Glo-1, and Tpi. The data from these enzymes have yet to be summarized.

Discussion

Sheppard and Berlocher (1989) stated that "electrophoresis will be a powerful tool for the discrimination of disputed species in *Apis*, since it so readily differentiated the species we examined." Their prediction was upheld for the two newly recognized species, *A. koschevnikovi* and *A. andreniformis*, which have unique patterns that differentiate them from all other honey bee species. For example, the two sister species *koschevnikovi* and *cerana* differ in the expression of Est-1 or Est-2 and in their Mdh, Fum, and Sudh alleles. The other two sister species, *florea* and *andreniformis*, share no Est-2 or Sudh alleles and are fixed for different Mdh alleles; and although they share allozymes for Fum, the allozyme frequencies are very different. There were striking differences in the patterns of allozyme variation in all of the seven taxa we studied except the two morphotypes of giant honey bees (see below). Preliminary estimates of genetic distance indicate that there are many differences between these taxa, as was also indicated by Sheppard and Berlocher (1989) for the four traditional species of *Apis*.

Because of these striking differences in allozyme patterns in different species of honey bees, Sheppard and Berlocher (1989) went on to suggest that "the resolution of the *A. dorsata* 'complex' seems imminently solvable by electrophoresis." In our study we analyzed both typical *dorsata* and the morphotype occurring on Sulawesi, *A. binghami* Cockerell. Not only did these two morphotypes have all allozymes in common (with the minor exception of two rare Mdh alleles), but the frequencies of the allozymes at each locus were similar as well. Therefore, the allozyme data do not support the species status of the giant honey bees of Sulawesi. The lack of any male genitalic differences between *dorsata* and *binghami* (G. Koeniger, pers. comm.) also support this tentative conclusion. This is somewhat surprising given the differences in color pattern and morphology between *dorsata* and *binghami* (Maa, 1953), differences in nesting strategies (Chapter 2, this volume), and probable isolation of Sulawesi from the rest of Southeast Asia at various times in geological history (Audley-Charles, 1981). Moreover, D. R. Smith has found mtDNA differences between these two morphotypes (Chapter 8, this volume). It would appear that mtDNA and morphological characters have diverged more rapidly than have allozymes. In the absence of more conclusive data, a conservative conclusion would be to accept *binghami* as no more than a distinct subspecies of *Apis dorsata*.

The most striking aspect of our study was the surprising number of polymorphic loci that we detected. A summary of the observed heterozygosities for all species and loci (Table 7.3) indicates the high degree of polymorphisms we recorded. Four of these isozymes have not been recorded previously for honey bees: α-Gdh, Fum, Sudh, and Shdh. For the ten loci summarized in Table 7.3, the average heterozygosity was 0.137. Of the other seven isozymes we typed, two (Idh and Me) were also polymorphic, so the average heterozygosity for all 17 loci over all species is at least 0.08-0.10. Previous estimates of heterozygosity for *Apis mellifera* range from 0.01-0.02 (summarized by Sylvester, 1986). The only estimates for Asian species of *Apis* are 0.002 for *cerana*, 0.003 for *dorsata*, and 0.023 for *florea* (expected heterozygosities calculated from allozymes frequencies given in Sheppard and Berlocher, 1989).

There are several possible reasons for these unusually high levels of genetic diversity for eusocial hymenopterans. First, our choice of enzymes may have been highly fortuitous. However, several of the isozymes we found to be polymorphic (α-Gdh, Fum, Sudh, Idh) have been typed for *Apis mellifera* previously without indications of polymorphism. Second, our values for heterozygosity have been calculated for each species, which combines within population and interpopulation variation. However, this does not appear to explain our high levels of heterozygosity either, because the most diverse species we recorded, *mellifera*, came from only two populations (Guelph, Canada, and Australia), with relatively small numbers of bees and colonies being represented. Even in those species for which only a single population was analyzed (*florea, koschevnikovi, binghami*), the average heterozygosity was 3 to 10 times greater than has been reported for *A. mellifera* in the past. Moreover, when a species was collected from more than one location, the allozymes did not notably vary between locations. Third, in contrast to other electrophoretic studies of Asian bees, our analyses were performed on species from near the center of their ranges where one might expect higher diversity. However, this again fails to explain the high diversity we recorded for *mellifera*. Fourth, we may have screened more electrode and gel buffer systems which allowed better detection of allozymes; there is no way to evaluate whether this was a factor influencing our results. However, several researchers have independently and extensively analyzed *Apis mellifera* without finding such high levels of polymorphism. Finally, we may have recorded variability where none actually exists. This is not likely because our results were consistent between trials and the banding patterns were clearly differentiable on the gels. It appears that we have discovered a wealth of genetic diversity which may be of substantial use to researchers studying either *mellifera* or the Asian honey bees.

Future Directions

Our results, although preliminary, suggest that honey bees have more allozyme variability than has been reported elsewhere. These results need to be reported in full by ourselves and verified by independent researchers. The relatively large number of polymorphic isozymes and shared alleles between species leaves open the possibility of fruitful application of clustering programs which will allow for phylogenetic and biogeographic inferences to be drawn.

The discovery of previously unreported polymorphisms for several isozymes is exciting. These should facilitate behavioral studies that use allozyme markers to identify different patrilines. The use of allozymes to identify Africanized bees, currently hindered by a lack of sufficient polymorphic isozymes (Daly, 1991), may be given a new boost if some of these isozymes differ between European and African bee races.

The *A. dorsata* complex represents a group of taxa which are clearly closely related yet have some distinct differences (Chapter 2, this volume). Because their distributions are allopatric, the species statuses of the bees in the Philippines (*A. breviligula*) and in Sulawesi (*A. binghami*) are still open to question. Allozyme studies are often useful in detecting cryptic species. Our results suggest that *A. binghami* Cockerell should be considered a subspecies of *A. dorsata*. Additional isozyme analyses of *A. laboriosa, A. breviligula, A. dorsata* from the Andaman and Nicobar Islands, and *A. dorsata* from southern India would be helpful for further understanding the taxonomic statuses of these taxa.

These suggested studies are the obvious spin-offs of our discoveries. It is difficult to predict other ways in which these results may affect further research. If our results prove to be correct, however, we will soon enter a new era of research on honey bee allozymes.

Literature Cited

Audley-Charles, M. G. 1981. Geological history of the region of Wallace's line. In *Wallace's Line and Plate Tectonics.*, T. C. Whitmore, ed. Oxford University Press: Oxford. Pp. 24-35.

Avise, J. C. 1974. Systematic value of electrophoretic data. *Syst. Zool.* 23:465-481.

Berlocher, S. H. 1984. Insect molecular systematics. *Annu. Rev. Entomol.* 29:403-433.

Buth, D. G. 1984. The application of electrophoretic data in systematic studies. *Annu. Rev. Ecol. Syst.* 15:501-522.

Cornuet, J. M. 1979. The MDH system in the honeybees (*Apis mellifera* L.) of Guadaloupe. *J. Hered.* 70:223-224.

Daly, H. V. 1991. Systematics and identification of Africanized honey bees. In *The "African" Honey Bee.*, M. Spivak, D. J. C. Fletcher and M. D. Breed, eds. Westview Press: Boulder, CO. Pp. 13-44.

Eldridge, B. F., Munstermann, L. E. and Craig, G. B. Jr. 1986. Enzyme variation in some mosquito species related to *Aedes (Ochlerotatus) stimulans* (Diptera: Culidicae). *J. Med. Entomol.* 23:423-428.

Harris, H. and D. A. Hopkinson. 1976. *Handbook of enzyme electrophoresis in human genetics.* North Holland Publishing Co.: Oxford.

Hartl, D. L. 1988. *A Primer of Population Genetics.*, 2nd Ed. Sinauer Associates Inc.: Sunderland, MA.

Hillis, D. M. 1987. Molecular versus morphological approaches to systematics. *Annu. Rev. Ecol. Syst.* 18:23-42.

Li, S., Y. Meng, J. T. Chang, J. Li, S. He, and B. Kuang. 1986. A comparative study of esterase isozymes in 6 species of *Apis* and 9 genera of Apoidea. *J. Apic. Res.* 25:129-133.

Maa, T. C. 1953. An inquiry into the systematics of the tribus Apidini or honeybees (Hymenoptera). *Treubia* 21:525-640.

Menken, S. B. J. 1980. *Allozyme polymorhism and the speciation process in small ermine moths (Lepidoptera, Yponomeutidae). (Studies in Yponomeutidae 2).* Ph. D. dissertation, University of Leiden, The Netherlands.

Menken, S. B. J. 1982. Biochemical genetics and systematics of small ermine moths (Lepidoptera, Yponomeutidae). *Z. zool. Syst. Evolutionforsch.* 20:131-143.

Metcalf, R. A., J. C. Marlin and G. S. Whitt. 1975. Low levels of genetic heterozygosity in Hymenoptera. *Nature* 257:792-794.

Munstermann, L. E. 1979. *Isozymes of Aedes aegypti: Phenotype, linkage and use in the genetic analysis of sympatric subspecies populations in East Africa.* Ph. D. Dissertation, University of Notre Dame, USA.

Nunamaker, R. A., W. T. Wilson and R. Ahmad. 1984a. Malate dehydrogenase and non-specific esterase isoenzymes of *Apis florea, A. dorsata,* and *A. cerana* as detected by isoelectric focusing. *J. Kans. Entomol. Soc.* 57:591-595.

Nunamaker, R. A., W. T. Wilson, and B. E. Haley. 1984b. Electrophoretic detection of Africanized honey bees (*Apis mellifera scutellata*) in Guatemala and Mexico based on malate dehydrogenase allozyme patterns. *J. Kans. Entomol. Soc.* 57:622-631.

Pamilo, P., S. Varvio-Aho and A. Pekkarinen. 1978. Low enzyme gene variability in Hymenoptera as a consequence of haplodiploidy. *Hereditas* 88:93-99.

Robinson, G. E. and R. E. Page Jr. 1988. Genetic determination of guarding and undertaking in honey-bee colonies. *Nature* 333:356-358.

------. 1989. Genetic determination of nectar foraging, pollen foraging, and nest-site scouting in honey bee colonies. *Behav. Ecol. Sociobiol.* 24:317-323.

Robinson, G. E., R. E. Page Jr. and M. K. Fondrk. 1990. Intracolonial behavioral variation in worker oviposition, oophagy, and larval care in queenless honey bee colonies. *Behav. Ecol. Sociobiol.* 26:315-323.

Ruttner, F. 1988. *Biogeography and Taxonomy of Honeybees.* Springer-Verlag: Berlin.

Shaw, C. R. and R. Prasad. 1970. Starch gel electrophoresis of enzymes -- a compilation of recipes. *Biochem. Genet.* 4:297-320.

Sheppard, W. S. 1988. Comparative study of enzyme polymorphism in United States and European honey bee (Hymenoptera: Apidae) populations. *Ann. Entomol. Soc. Am.* 81:886-889.

Sheppard, W. S. and S. H. Berlocher. 1989. Allozyme variation and differentiation among four *Apis* species. *Apidologie* 20:419-431.

Sheppard, W. S. and M. D. Huettel. 1988. Biochemical genetic markers, intraspecific variation, and population genetics of the honey bee. In *Africanized Honey Bees and Bee Mites.*, G. R. Needham, R. E. Page, Jr., M. Delfinado-Baker and C. E. Bowman, eds. Ellis-Horwood, Ltd.: Chichester, England. Pp. 281-286.

Sheppard, W. S. and B. A. McPheron. 1986. Genetic variation in honey bees from an area of racial hybridization in western Czechoslovakia. *Apidologie* 17:21-31.

Spencer, N., D. A. Hopkinson and H. Harris. 1964. Phosphoglucomutase polymorphism in man. *Nature* 204:742-745.

Steiner, W. W. M. and D. J. Joslyn 1979. Electrophoretic techniques for the genetic study of mosquitoes. *Mosquito News* 39:35-39.

Sylvester, H. A. 1982. Electrophoretic identification of Africanized honeybees. *J. Apic. Res.* 21:93-97.

------. 1986. Biochemical genetics. In *Bee Genetics and Breeding.*, T. E. Rinderer, ed. Academic Press: Orlando, FL. Pp. 177-203.

Tanabe, Y. and Y. Tamaki. 1985. Biochemical genetic studies on *Apis mellifera* and *Apis cerana*. Proc. 30th International Beekeeping Congress, Nagoya, Japan. Pp. 152-154.

Varvio-Aho, S. and P. Pamilo. 1980. A new buffer system with wide applicability. *Isozyme Bulletin* 13:114.

8

Mitochondrial DNA and Honey Bee Biogeography

Deborah Roan Smith

It may seem that there is little connection between laboratory studies of mitochondrial DNA molecules and the biogeography of free-living honey bee populations. However animal mitochondrial DNA (mtDNA) has a number of properties that make it a favorite tool of systematists and population biologists, and study of mtDNA has given biologists new insights into the systematics, biogeography and population biology of many taxa, including the genus *Apis*.

Apis species can be grouped on the basis of morphology and behavior into three lineages: the cavity-nesting bees, *A. mellifera, A. cerana* and *A. koschevnikovi*; the dwarf bees, *A. florea* and *A. andreniformis*; and the giant bees, *A. dorsata*. Each of these lineages occurs over a wide range of habitats and climates. Each of the three lineages consists of numerous morphologically differentiated populations ranging from ecotypes and allopatric subspecies to reproductively isolated sympatric species, and many populations, such as those on islands in the Indonesian, Malaysian and Philippine archipelagos, are geographically isolated. As a result, the nomenclature and systematics of *Apis* are complex. At one extreme, Maa (1953) divided the honey bees into three genera: *Micrapis*, the dwarf bees, with two species; *Megapis*, the giant bees, with four species; and *Apis*, the cavity-nesting bees. *Apis* was further subdivided into two genera, *Apis* (*Apis*), the European and African cavity-nesting bees, with seven species and several subspecies; and *Apis* (*Sigmatapis*), the Asian cavity-nesting bees, with 11 species and several subspecies. Subsequent workers tended to ignore Maa's classification and recognized only four species of honey bees in the genus *Apis*, one species corresponding to each of Maa's genera or subgenera: *Apis florea*, the dwarf bee, *A. dorsata*, the giant bee, *A. cerana* the eastern cavity-nesting bee, and *A. mellifera*, the western cavity-nesting bee.

Recent work on the Asian bees (summarized in Chapter 2, this volume) has led to the "rediscovery" of two of Maa's species, *A. andreniformis* (Wu and Kuang, 1987; Wongsiri et al., 1990) and *A. koschevnikovi* (Koeniger et al., 1988; Tingek et al., 1988; Rinderer et al., 1989; Ruttner et al., 1989). This has renewed interest in Maa's classification; although few are willing to accept three genera and 24 species of honey bees, it is clear that many of the populations he recognized as species deserve reexamination.

The present study of mtDNA diversity in *Apis* addresses five main questions:

1. How many distinct mitochondrial haplotypes or groups of related haplotypes are found in each of the *Apis* species now recognized?
2. What are the geographic distributions of different mtDNA haplotypes within each species?
3. Do mtDNA data support the validity of species and subspecies recognized on morphological grounds?
4. What are the phylogenetic relationships among mitochondrial haplotypes?
5. The ranges of the three *Apis* lineages overlap extensively in India and southeast Asia; are the geographic distributions of mitochondrial haplotypes in this region the same or different in each lineage? Do the patterns of mtDNA differentiation suggest congruent patterns of colonization and differentiation for the three *Apis* lineages, or idiosyncratic histories?

Before presenting the results of this study, I discuss the rationale for using mitochondrial DNA in evolutionary studies and summarize the extant literature on *Apis* mitochondrial DNA.

Why Mitochondrial DNA?

Mitochondria, the organelles responsible for aerobic metabolism in eucaryotic cells, possess their own small DNA molecules or chromosomes, distinct from the chromosomes of the cell nucleus. Each mitochondrion contains several copies of its chromosome, and each cell may contain one to hundreds of mitochondria. Thus the first of the useful properties of animal mtDNA is its high copy number: while any given nuclear gene may be present in just two copies per cell (one from each parent in diploid organisms) the mitochondrial genome is present in many copies per cell. Some metabolically active tissues, such as eggs or insect flight muscle, are particularly rich in mitochondria and mtDNA. High copy number makes it easy to isolate and purify enough mtDNA for many types of analysis directly from animal tissues, without the necessity of isolating and cloning individual genes.

Another useful characteristic of animal mtDNA is its small size. Most animal mitochondrial genomes consist of a single small, circular molecule. The animal mitochondrial chromosomes which have been isolated and studied so far range in size from 14,284 base pairs (bp) in the nematode *Ascaris suum* (Wolstenholme et al., 1987) and approximately 14,000 bp in the scorpion *Hadrurus arizonensis* (Smith and Brown, in press), to 32,000 bp in the bark beetles in the genus *Pissodes* (Boyce et al., 1989) and 39,000 or more bp in the sea-scallop *Placopecten magellanicus* (Snyder et al., 1987). These mitochondrial genomes are several orders of magnitude smaller than nuclear genomes, and consequently the entire mitochondrial genome can be studied as a unit.

A third useful property of animal mtDNA is its generally conservative gene content and gene arrangement. Most animal mtDNAs code for only 12 or 13 proteins, two ribosomal RNAs (rRNAs), and 22 transfer RNAs (tRNAs) (Brown, 1985; Wolstenholme et al., 1987). The order or placement of these genes on the mtDNA molecule, particularly the protein and rRNA genes, is quite stable within major taxa (Moritz et al., 1987).

Complete and partial gene orders have been obtained for many insect mitochondrial genomes, among them the fruit flies *Drosophila yakuba* (Clary and Wolstenholme, 1985) and *D. melanogaster* (De Bruijn, 1983; Garesse, 1988), the mosquitos *Aedes albopictus* (HsuChen and Dubin, 1984; HsuChen et al., 1984; Dubin et al., 1986) and *Anopheles quadrimaculatus* (Cockburn et al., 1990), the locust *Locusta migratoria* (McCracken et al., 1987; Uhlenbusch et al., 1987; Haucke and Gellissen, 1988) and the bark beetles *Pissodes* (Boyce et al., 1989). These insect mtDNAs show many rearrangements of protein rRNA and tRNA genes relative to the vertebrate gene order. Among insects, the genes for proteins and rRNAs have been found to be in the same relative positions as in *D. yakuba*; when rearrangements have been found, they involve the numerous, small, structurally similar tRNA genes. Within species and genera and perhaps even families, gene order is expected to be uniform, with the exception of novel mutations producing duplications or inversions of sections of the genome. This means that it is relatively easy to homologize or "line up" and compare stretches of mtDNA from related populations and species. Comparisons among animal mitochondrial genomes are made even easier by the fact that, unlike the nuclear genome, the typical animal mtDNA does not contain either introns or large spacers between genes (Brown, 1985).

The rate of evolution of a gene or other piece of DNA is also a very important factor determining its usefulness in systematic or biogeographic studies. When two lineages diverge, each begins to accumulate mutations independently of the other. Early in the process, it is unlikely that a mutation will occur at the same base pair as a previous mutation; however over time, as mutations in each lineage accumulate, it becomes more and more probable that a mutation will take place at a DNA base pair that has already sustained a mutation. These "multiple hits" obscure the historical information recorded in the DNA sequence. Thus a gene or other segment of DNA is phylogenetically informative in a comparison of taxa if the gene is still in the early phase of divergence, while the probability of multiple hits is still low (Brown et al., 1982). Poorly-conserved, rapidly-evolving genes are useful in comparisons of recently diverged lineages or taxa, while highly-conserved, slowly-evolving genes are useful for comparison of more ancient taxa.

Mitochondrial DNA is useful in phylogenetic studies over a range of divergence times, since different portions of the mitochondrial genome evolve at different rates. Simon (1991) has reviewed information on the relative rates of evolution of genes in the mitochondrial genome. Briefly, the non-coding A+T rich region of invertebrate mtDNA evolves rapidly, while genes such as those coding for the large and small rRNA, or cytochrome oxidase subunits I and II (COI and COII) evolve very slowly. The phylogenetic "reach" of protein-coding genes can be increased by translating the DNA sequence into amino acid sequence, which reduces the effect of the more rapidly accumulated silent base pair substitutions, i.e., those that do not result in an change in amino acid.

A final set of properties are particularly useful for the population biologist: animal mtDNA is typically (though not invariably) maternally inherited without recombination (Lansman et al., 1983; Brown, 1985; Gyllensten et al., 1985; but see Satta et al., 1988; Kondo et al., 1990; Hoeh et al., 1991). Only maternal inheritance of mtDNA has been demonstrated for honey bees (Meusel and Moritz, 1990). This has several implications for the study of honey bee mtDNA. First, all the offspring of a queen carry the same mitochondrial DNA; this means that large quantitites of a particular mitochondrial genome can be prepared by pooling mtDNA from workers in a hive. Second, where strict maternal inheritance is the rule, individuals of hybrid origin do not carry a mixture of parental mtDNAs; they show only the mother's mtDNA. This is particularly useful when studying the origins and population biology of hybrid populations (e.g., Hall and Muralidharan, 1989; Smith et al., 1989).

Apis Mitochondrial DNA

As is true of so many aspects of genetics, more is known about the mitochondrial genome of *Drosophila* than any other insect. Surprisingly, the mitochondrial genome of *Apis* may soon be the second best-studied in the insect world. The current interest in honey bee mitochondrial DNA can probably be traced to two sources: the study of genetic differentiation among *A. mellifera* subspecies and the search for genetic markers to monitor the spread of African and Africanized honey bees in the Americas.

Apis mellifera mtDNA

The study of genetic differentiation among *Apis mellifera* subspecies is a logical continuation and extension of the numerous studies of morphological, ecological and behavioral differentiation among these populations (summarized in Ruttner, 1988). Honey bees were among the first organisms examined for allozyme polymorphisms using protein electrophoresis (e.g., Mestriner, 1969; Mestriner and Contel, 1972; Brückner, 1974). Unfortunately, little variation was detected, particularly in the earliest studies. The first published account of mtDNA polymorphisms in honey bee mtDNA (Moritz et al., 1986) was a comparison of mitochondrial restriction fragment length polymorphisms (RFLPs: see Methods and Materials, below) in Australian honey bees derived from three European subspecies: *A. m. carnica, ligustica* and *caucasica*. This study revealed only one polymorphism among these three subspecies. Subsequent surveys of mtDNA polymorphisms within and among *A. mellifera* subspecies (Smith, 1988, 1991; Smith and Brown 1988, 1990; Smith et al., 1991), discussed in more detail below, showed that *A. mellifera* mtDNA has both restriction site polymorphisms and size polymorphisms.

MtDNA polymorphisms have been used extensively in population studies of Neotropical African and Africanized honey bees in the Americas (e.g., Hall and Muralidharan, 1989; Smith et al., 1989; Y. Crozier et al., 1991; Hall and Smith, 1991; Sheppard et al., 1991).

Portions of the *A. mellifera* mitochondrial genome have been sequenced: the sequence of most of the large subunit ribosomal RNA gene (lrRNA), the cytochrome oxidase subunits I and II (COI and COII) and several tRNA genes have been published (Vlasak et al., 1987; R. Crozier et al., 1989). Like the mtDNA of *Drosophila yakuba* (Clary and Wolstenholme, 1985) and other insects that have been examined, the mtDNA of *A. mellifera* has a very high proportion of the bases adenine and thymine (it is "A+T rich"); the *A. mellifera* COI and COII genes have an even higher percentage of A+T than those of *D. yakuba* (R. Crozier et al., 1989). Published and unpublished work indicates that the genes for the small subunit ribosomal RNA (srRNA), lrRNA, COI, COII, COIII, cytochrome b, and ATPase 6 and 8 are in the same relative positions as in *D. yakuba*, although there are rearrangements of the tRNA genes (R. Crozier et al., 1989; Y. Crozier et al., in press). Comparison of sequence data from vertebrate, *D. yakuba* and *A. mellifera* mitochondrial genes suggests that *A. mellifera* mitochondrial DNA (and perhaps all *Apis* mtDNA) is evolving at an unusually high rate (R. Crozier et al., 1989).

The mitochondrial DNA molecule of *Apis mellifera* contains 16,500 to 17,000 base pairs (Smith and Brown, 1988; Smith and Brown, 1990). The size of the *A. mellifera* mtDNA molecule differs among populations and among individuals. Some elements of size variation appear to be due to small, discrete pieces of DNA which may be repeated a variable number of times. The locations of two such elements have been mapped. A small element of approximately 80 bp is present in one to three copies in *A. m. mellifera* (Smith and Brown, 1990). A larger repeating element is found in honey bees of the west European lineage (*A. m. mellifera* and some *A. m. iberica*) and African lineage (including some *A. m. iberica*, *A. m. intermissa* and *A. m. scutellata*) (Smith 1988; Smith and Brown, 1988, 1990; Smith et al., 1991), giving rise to five size classes (Hall and Smith, 1991). This element lies between the COI and tRNA^{leu} genes (5′) and the COII gene (3′); only 5 bp are found at this spot in the mtDNA of *D. yakuba* (Clary and Wolstenholme, 1985; R. Crozier et al., 1989). This region has recently been sequenced (Cornuet et al., 1991). It consists of two units, a "P" unit of 54 bp which is 100% A+T, and a "Q" unit of 196 bp, which shows sequence similarity to the 3′ end of the COI gene (the "Q1" portion), the tRNA^{leu} gene (the "Q2" portion) and the "P" sequence (the "Q3" portion). The size classes correspond to the combinations: Q, PQ, PQQ, and PQQQ, with the Q units in direct tandem repeats. It seems likely that these repeats arose as a result of a duplication of portions of the COI and the tRNA^{leu} gene, followed by additional tandem duplications (Cornuet et al., 1991).

Interspecies Comparisons.

Apis mtDNA is a rich source of phylogenetic information. In addition to the studies by Cameron (Chapter 4, this volume) and Sheppard and McPheron (Chapter 5, this volume), Garnery et al. (1991) have sequenced portions of the COII gene in *A. mellifera, cerana, dorsata* and *florea*. Sequence divergence among the *Apis* species ranged from approximately 7% to 11%. Garnery et al. (1991) found that *A. mellifera* and *A. cerana* were more closely related to one another than to the other two species. Branching order among the *dorsata, florea* and *mellifera+cerana* lineages could not be determined.

The intergenic region between COI, tRNA[leu] and COII is another interesting source of phylogenetic information. Portions of this region were sequenced in *A. cerana, A. dorsata, A. florea, Bombus leucorum* (Apidae) and *Xylocopa violacea* (Anthophoridae) (Cornuet et al., 1991). All four of the *Apis* species examined have an intergenic region which is A+T rich; in *A. cerana* and *A. m. mellifera* (*A. koschevnikovi* was not examined) there is the additional intergenic region with sequence homology to COI and tRNA[leu] discussed above. Both these intergenic regions are absent in *B. leucorum* or *X. violacea*.

These data suggest that this intergenic region may be an autapomorphy for Apini. However this region has not been examined in Euglossines or Meliponines, and could equally well prove to be a synapomorphy for Apini+Euglossini (a clade suggested by the work in Chapter 4, this volume) or for Apini+Meliponini (a clade suggested by the work in Chapter 3, this volume).

Methods and Materials

General information on methods are given here; methods particular to each species are given below along with results. In most cases, 25-100 adult workers were collected directly from colonies; in a few cases individual workers were collected from flowers. The bees were frozen in liquid nitrogen or on dry ice and transported to the University of Michigan Laboratory for Molecular Systematics, where they were stored at -80° C until used in DNA preparation. Information on collection sites and samples sizes is presented in the Appendix.

Analysis of Mitochondrial DNA

There are many ways for the biologist to extract information from DNA, but two types of data in particular are relevant to the study of honey bee biogeography and systematics: sequence information and restriction site information, each of which has advantages and disadvantages. The sequence or listing of the actual DNA base pairs that make up a particular gene or DNA molecule provides the most complete type of information. Because of the labor involved only a few animal mitochondrial genomes have been sequenced in their entirety: two nematodes, *Ascaris suum* and *Caenorhabditis elegans* (Wolstenholme et al., 1987); a fly, *Drosophila yakuba* (Clary and Wolstenholme, 1985); the sea urchins *Strongylocentrotus purpuratus* (Jacobs et al., 1988) and *Paracentrotus lividus* (Cantatore et al., 1989); and several vertebrates (Anderson et al., 1981; Bibb et al., 1981; Anderson et al., 1982; Roe et al., 1985; Desjardins and Morais, 1990). Several other mitochondrial genomes have been partially sequenced, including those of honey bees (see below).

It is impractical to sequence the entire mitochondrial genome in studies involving many different taxa, or many individuals within a population. Instead, it is necessary to select small portions of the mitochondrial genome to compare among taxa or among individuals. The recent exploitation of the polymerase chain reaction (PCR: Saiki et al., 1985) to amplify selected stretches of DNA for sequencing has made this approach very attractive (Simon et al., 1991; Chapters 4 and 5, this volume). Sequencing of PCR amplified DNA is used to provide detailed information about relatively short stretches of DNA. One problem with this approach is ensuring that there is enough information in the piece of DNA selected to answer a particular question. Highly-conserved, slowly-evolving regions in the mitochndrial genome may not have enough variation to provide information on the phylogenetic relatedness of closely related taxa or populations: all the genomes examined may be essentially identical. On the other hand, in poorly-conserved regions there may be very high levels of variation, suitable for distinguishing individual family lines in a small population. However, such regions would not be useful in comparing distantly related populations or taxa: many individual base pairs will have undergone multiple mutations, resulting in high levels of homoplasy.

An alternative is to use restriction endonucleases to compare the mitochondrial genomes of individuals or taxa. This approach is relatively simple and provides information from all regions of the mitochondrial genome. It is a way to scan the entire mitochondrial genome, but, of course, it does not provide the detailed information of sequence data.

Restriction endonucleases, or restriction enzymes, bind to specific sequences of DNA base pairs and cut the DNA strand. Most commercially available restriction enzymes recognize particular sequences of four, five or six base pairs (4-base, 5-base or 6-base enzymes). In the example in Figure 8.1, samples of *Apis mellifera ligustica* and *A. m. scutellata* mtDNA are digested with the restriction enzyme *Eco*R I, which recognizes sequence GAATTC. There are four sites with this sequence in *A. m. ligustica* mtDNA; *Eco*R I cleaves the circular mtDNA at these four restriction sites yielding four pieces, or restriction fragments. The mtDNA of an individual from another population or species may have more or fewer *Eco*R I restriction sites. *A. m. scutellata* mtDNA has only three; one restriction site which is present in *A. m. ligustica* mtDNA is not present in *A. m. scutellata* mtDNA. This variation is observed as differences in the length of the restriction fragments generated by the restriction enzyme, and is referred to as restriction fragment length polymorphism (RFLP).

While each restriction enzyme recognizes only one DNA sequence, there are many dozens of restriction enzymes with different recognition sites available commercially. Different mitochondrial genomes can be compared simply by comparing the number and size of restriction fragments generated by each restriction enzyme. In a more labor-intensive study, each restriction site can be mapped on the mitochondrial DNA molecule, resulting in a *restriction site map* like the ones shown in Figures 6.3, 6.4 and 8.2. Mapped restriction sites can be treated as discrete two-state characters (present or absent) in a cladistic analysis. (Cladistic analysis of RFLPs is also possible, but homoplasy presents a more serious difficulty in fragment comparisons).

It is often useful to have a statistic which simply describes the differences between two DNA molecules, such as the mitochondrial DNA molecules from two *Apis mellifera* subspecies; the most commonly used statistic is percent sequence divergence (e.g., Nei and Tajima, 1983). This is an estimate of the probability that a particular base pair has undergone a mutation since two lineages diverged, or alternatively, an estimate of the percentage of base pairs in a DNA molecule that have undergone mutation since two lineages diverged. Percent sequence divergence among mitochondrial genomes may be estimated from RFLP data or from mapped restriction site data (Nei, 1987), however estimates of sequence divergence from RFLP data are unreliable for divergences greater than approximately 5%.

140

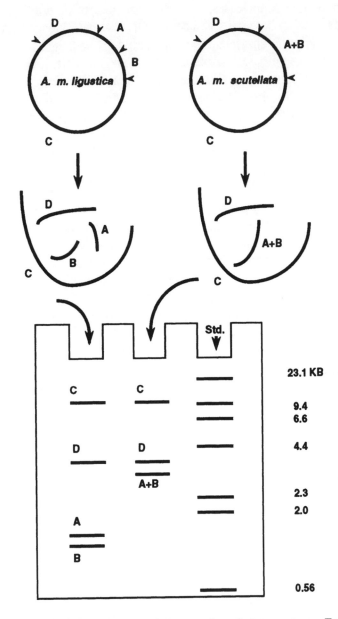

Figure 8.1 Schematic explanation of the use of restriction enzymes. Top panel: Mitochondrial DNAs of *Apis mellifera ligustica* (left) and *A. m. scutellata* (right) are digested with the restriction enzyme *Eco*R I. *A. m. ligustica* has four *Eco*R I recognition sites, *A. m. scutellata* has three. Middle panel: the mtDNA of *A. m. ligustica* is cleaved into four fragments (A, B, C, D) and that of *A. m. scutellata* into three fragments (A+B, C, D). Bottom panel: the resulting fragments are separated by gel electrophoresis along with a size standard.

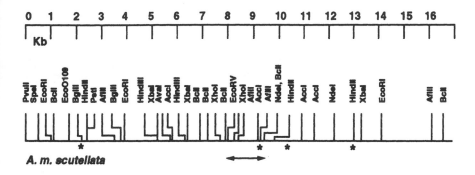

Figure 8.2 Restriction site map of the mtDNA of *A. mellifera scutellata*. Asterisks indicate sites that may be present or absent in different *A. m. scutellata* haplotypes. The double-headed arrow indicates the location of the element of size variation discussed in Table 8.3.

Sample Preparation and Analysis

Mitochondrial DNA was prepared from four to six colony-mates per sample of *A. mellifera, cerana, koschevnikovi* and *dorsata*, and from 10-12 colony-mates per sample of *A. florea* and *andreniformis*, using previously described methods (Smith and Brown, 1990; Smith et al. 1991). The Asian samples were surveyed with five or six restriction enzymes having six-base recognition sequences (6-base enzymes). *Apis mellifera* samples were surveyed with 15 or 16 6-base restriction enzymes. In all cases the temperature and buffer conditions were those recommended by the manufacturers. Restriction fragments were radioactively labeled with ^{32}P nucleotides using DNA polymerase I (large fragment) and separated by electrophoresis on 1% or 1.5% agarose gels and 4% polyacrylamide gels. The restriction fragments were visualized on X-ray film by autoradiography.

The mtDNA of *A. mellifera* has been studied in greatest detail, and restriction site maps have been constructed for each subspecies examined. The mapped restriction sites were used to estimate percent sequence divergence among mitochondrial genomes of *A. mellifera* subspecies following methods of Nei and Tajima (1983). Study of the mtDNA of Asian *Apis* populations is in a much earlier stage; to date only restriction fragment length polymorphisms have been used to identify mitochondrial lineages within each species. Percent sequence divergence among mitochondrial genomes of the Asian bees was estimated from shared restriction fragment data following methods described by Nei (1987). Two programs written by R. Hagen for PC-type microcomputers were used to calculate sequence divergences (R. H. Hagen, pers. comm.).

Distance phenograms for the mitochondrial haplotypes of each species were constructed from the estimates of sequence divergence between each pair of unique haplotypes using UPGMA (Sneath and Sokal, 1973). These phenograms indicate major mitochondrial lineages or clusters within species.

Data from mtDNA restriction enzyme cleavage site maps and the taxonomic distribution of mtDNA size polymorphisms provides information on the phylogenetic relationships of *A. mellifera* subspecies. Mapped restriction sites were treated as unweighted characters and a cladistic analysis was carried out using Hennig86, a computer program by James S. Farris (Farris, 1988). At this point only RFLP data exist for the Asian populations; although these data provide information on the number of distinct lineages in the Asian species, analysis of phylogenetic relationships of these populations and species will require mapped restriction site and sequence data.

Analysis and Results

Apis mellifera

Background. *Apis mellifera*, the western cavity-nesting bee, occurs over a wide range of climates and habitats in Europe, Africa and the Middle East. It has been subdivided into numerous geographic races, or subspecies, on the basis of ecological, behavioral and morphological differences between populations (Adam, 1983; Ruttner, 1988 and references therein). Ruttner (1988) has reviewed the literature on *A. mellifera* subspecies, and recognized 24 subspecies (see Ruttner, 1988 for the natural ranges of these subspecies). Principal component analyses of morphometric measurements of *A. mellifera* subspecies indicated three (Ruttner et al., 1978) or four (Ruttner, 1988) major clusters: a western Mediterranean cluster (the "M-group"), which includes the north and west European black bee, *A. mellifera mellifera*, the bee of Spain and Portugal, *A. m. iberica*, north African *A. m. intermissa*, and several other north African subspecies; a tropical African cluster (the "A-group"), which includes *A. m. lamarckii* from Egypt, *A. m. scutellata*, *A. m. capensis* and other sub-Saharan subspecies, and *A. m. unicolor* from Madagascar; and an eastern Mediterranean cluster. The latter has been subdivided into a "C-group" which includes the Carniolan bee, *A. m. carnica* and the Italian bee, *A. m. ligustica*, and an "O-group" which includes *A. m. caucasica* and subspecies of the Middle East (Ruttner, 1988).

A. mellifera is usually considered a young species and its many subspecies are thought to have originated very recently in the late Pleistocene. The phylogenetic relationships of the three cavity-nesting honey bees, *A. mellifera, A. cerana* and *A. koschevnikovi* are not completely agreed upon, but *A. cerana* (or one of its subspecies) is usually believed to be the sister taxon to *A. mellifera*. The ranges of the two reach their closest approximation in the region of Iran, Pakistan and Afghanistan, and it has been suggested that the separation of *mellifera* and *cerana* from an ancestral population occurred here:

> The desert belt of Iran, unchanged since Oligocene, could have been the prerequisite for the separation of the sister species *cerana-mellifera*. Supposing an advance to the west, north of the Kavir, of an *Apis* population of the Indian subcontinent, the connection with the main area would necessarily been very narrow. An interruption of this connection by a climatic change could have resulted in complete isolation, creating the possibility of the origin of separate species (Ruttner et al., 1985).

Geographic differentiation of *A. mellifera* into numerous geographic races or subspecies can be attributed to colonization of the west by three or more routes, followed by differentiation of populations within each lineage during the Pleistocene (Ruttner, 1988; Ruttner et al., 1978). From its hypothesized origin in or near modern-day Iran, Ruttner (1968, 1988) hypothesized that one lineage ("C" + "O") colonized the Middle East and spread around the eastern shores of the Mediterranean; a second ("M") spread across North Africa and northwards into Europe around the western portion of the Mediterranean; and a third lineage ("A") colonized sub-Saharan Africa. Each of these lineages became further subdivided during the Pleistocene as populations became geographically isolated. As Ruttner has noted (Ruttner, 1968) during the Pleistocene glaciations most of Europe was uninhabitable for honey bees, and *A. mellifera* populations would have been confined to refugia in warmer Mediterranean peninsulas: the ancestors of *A. m. mellifera* and perhaps *A. m. iberica* in the Iberian peninsula and north Africa; *A. m. ligustica* in the Italian peninsula; and *A. m. carnica* in the Balkan peninsula. Expansion and contraction of forest and savannah habitat could have led to the geographic isolation of populations in Africa. For example, populations of *A. m. monticola* are now confined to forest in the higher elevations of mountains of East Africa; this disjunct range suggests that they once occupied a continuous region of humid forest (Meixner et al., 1989).

Several of the *A. mellifera* subspecies are isolated island populations: *A. m. adami* on Crete (Ruttner, 1980), *A. m. cypria* on Cyprus, and *A. m. sicula* on Sicily (Ruttner, 1988). The distribution of *A. m. unicolor* is

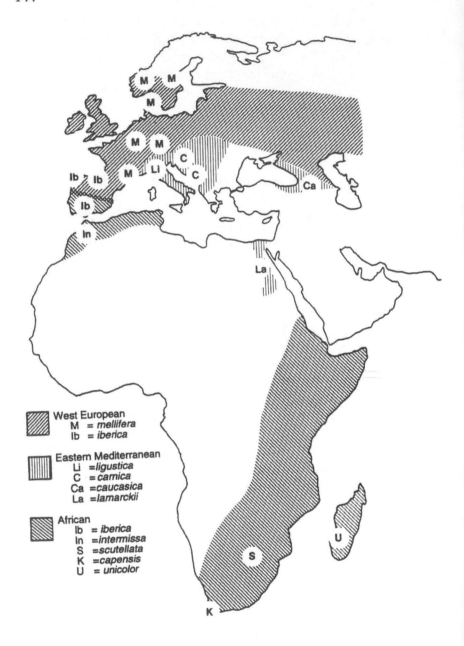

Figure 8.3 Dots indicate approximate location of collection sites for the *A. mellifera* used in this study. Shadings indicate the three lineages of mitochondrial haplotypes revealed by cluster analysis of sequence divergence among haplotypes (shadings are extended to the full range of the subspecies sampled).

particularly interesting. This subspecies is found on Madagascar, Réunion and Mauritius, but it was probably introduced to Réunion and Mauritius in the 17th and 18th century by French settlers from Madagascar (references in Ruttner, 1988). But, how these very black bees came to be in Madagascar is itself a problem. Because Madagascar separated from the African continent at least 70 to 90 million years ago (Owen, 1983), that is, long before the hypothesized origin of *A. mellifera*, it has always been considered unlikely that these bees were *in situ* before Madagascar separated from Africa. It has been suggested that they were imported from the coast of East Africa to Madagascar no more than 2,000 years ago by early Indonesian settlers (Anon., 1988). However, the very distinctive color and morphometric characters of this bee (Ruttner, 1988), and the fact that two ecotypes of the subspecies are found in tropical coastal areas and temperate central areas (Anon., 1988) suggests that honey bees colonized Madagascar much earlier, perhaps without the aid of humans.

MtDNA Analysis. For this study samples were obtained from 10 of the 24 *A. mellifera* subspecies recognized by Ruttner (1988). These include *A. m. mellifera* from Denmark, Norway, Sweden, France and Austria (the latter were "Nigra" hybrids); *A. m. carnica* from Germany, Austria, and Yugoslavia (transplanted to Austria); *A. m. ligustica* from Italy; *A. m. caucasica* from the Republic of Georgia, U.S.S.R. (transplanted to France); *A. m. iberica* from Spain; *A. m. lamarckii* from Egypt; *A. m. unicolor* from Madagascar; and *A. m. scutellata* and *A. m. capensis* from the Republic of South Africa. J.-M. Cornuet and L. Garnery shared information on *A. m. intermissa* from Morocco and Algeria (Chapter 6, this volume; Smith et al., 1991). More complete collection information is given in the Appendix. The approximate collection sites are shown in Figure 8.3.

MtDNAs prepared from these samples were digested with a set of 16 6-base restriction enzymes: *Acc* I, *Afl* II, *Ava* I, *Bcl* I, *Bgl* II, *EcoO* 109, *EcoR* I, *EcoR* V, *Hind* II, *Hind* III, *Nde* I, *Pst* I, *Pvu* II, *Spe* I, *Xba* I, and *Xho* I. The relative positions of restriction enzyme cleavage sites revealed by these enzymes were mapped by means of double digests. The resulting restriction site maps for most of these subspecies have been published elsewhere (Smith, 1988; Smith and Brown, 1988, 1990; Smith et al., 1991). These studies revealed two types of polymorphism in honey bee mtDNA: restriction site polymorphisms (that is, presence or absence of particular mapped restriction sites) and length variation.

The enzymes used here revealed a total of 48 restriction sites; 23 of these sites were found in all *A. mellifera* subspecies examined and 25 were polymorphic. The 25 polymorphic restriction sites distinguished 19

Table 8.1 Data matrix for parsimony analysis of the mitochondrial haplotypes found in *Apis mellifera* subspecies. The characters are 25 polymorphic restriction sites which can occur in the state "0" (absent) or "1" (present); "?" indicates missing data. The haplotypes of subspecies belonging to the eastern European group (lig1, car1, car2, car3 and lam1) were used as outgroups for the other haplotypes based on the character of the COI-tRNA^leu intergenic region (see text for discussion). Haplotypes: lig1 from *A. m. ligustica*; car1-3 from *A. m. carnica*; lam1 from *A. m. lamarckii*; mel1-4 from *A. m. mellifera*; int1-4 from *A. m. intermissa*; scu1-5 from *A. m. scuellata*; uni1 from *A. m. unicolor*. Approximate location of each restriction site on the *A. mellifera* mtDNA is given in kilobase pairs (kb) from the *Pvu* II site common to all subspecies (See Figure 8.2; Smith and Brown 1990).

MITOCHONDRIAL HAPLOTYPES

CHARACTERS		lig1	car1	car2	car3	lam1	mel1	mel2	mel3	mel4	int1	int2	int3	int4	scu1	scu2	scu3	scu4	scu5	uni1
00	*Acc* I, 10.5 kb	1	1	1	1	1	0	0	0	0	1	1	1	1	1	1	1	1	1	?
01	*Acc* I, 09.1 kb	0	0	0	0	0	0	0	0	0	0	0	0	0	0	0	0	0	0	0
02	*Afl* II, 09.1 kb	0	0	0	0	?	0	0	0	0	0	1	0	0	1	1	1	1	1	0
03	*Afl* II, 16.1 kb	0	0	0	0	?	0	0	0	0	0	0	0	0	1	1	1	1	1	1
04	*Ava* I, 05.6 kb	0	0	0	0	0	0	0	0	0	1	1	1	1	1	1	1	1	1	1
05	*Bcl* I, 06.8 kb	1	1	1	1	1	1	1	1	1	1	1	1	1	1	1	1	1	1	1
06	*Bgl* II, 00.5 kb	1	1	1	1	1	1	1	1	1	1	1	1	1	0	0	0	0	0	1
07	*Bgl* II, 07.2 kb	0	0	0	0	0	1	1	1	1	0	0	0	0	0	0	0	0	0	0
08	*Bgl* II, 02.2 kb	0	0	0	0	0	0	0	0	0	0	0	0	0	0	1	0	0	0	0
09	*EcoR* I, 02.5 kb	1	1	1	1	1	0	0	0	0	0	0	0	0	0	0	0	0	0	0
10	*EcoR* I, 14.8 kb	0	0	0	0	0	1	1	1	1	0	0	0	0	0	0	0	0	0	0
11	*Hind* II, 07.4 kb	0	0	0	0	0	1	1	1	1	0	0	0	0	0	0	0	0	0	0
12	*Hind* II, 09.7 kb	0	0	0	0	0	0	0	0	0	0	0	0	0	0	0	0	1	0	0

continues...

Table 8.1, *continued*

MITOCHONDRIAL HAPLOTYPES

CHARACTERS	lig1	car1	car2	car3	lam1	mel1	mel2	mel3	mel4	int1	int2	int3	int4	scu1	scu2	scu3	scu4	scu5	uni1
13 *Hind* II, 12.9 kb	0	0	0	0	0	0	0	0	0	0	0	0	0	1	0	1	0	0	0
14 *Hind*III, 06.3 kb	1	1	1	1	1	0	0	0	0	1	0	1	1	1	1	1	1	1	1
15 *Nde* I, 14.8 kb	1	1	1	1	1	1	1	1	1	1	0	0	0	0	0	0	0	0	0
16 *Nde* I, 15.7 kb	0	0	1	1	1	0	0	0	0	0	0	0	0	0	0	0	0	0	0
17 *Pst* I, 02.4 kb	1	1	1	1	0	1	1	1	1	1	1	1	1	1	1	1	1	1	1
18 *Spe* I, 00.5 kb	0	0	0	0	?	1	1	0	1	1	1	1	1	1	1	1	1	1	?
19 *Spe* I, 07.0 kb	1	1	1	1	?	0	0	0	0	0	0	0	0	0	0	0	0	0	?
20 *Spe* I, 15.1 kb	0	0	0	0	?	1	0	1	1	0	0	0	0	0	0	0	0	0	?
21 *Xba* I, 08.2 kb	0	0	1	1	0	0	0	0	0	0	0	0	0	0	0	0	0	0	0
22 *Xba* I, 08.3 kb	1	1	1	0	1	1	1	1	1	1	1	1	1	1	1	1	1	1	1
23 *Xba* I, 12.9 kb	1	1	1	0	0	1	1	1	1	1	1	1	1	1	1	1	1	1	1
24 *Xho* I, 08.2 kb	0	0	0	0	1	0	0	0	0	0	0	0	0	1	1	1	1	1	?

Table 8.2 Percent sequence divergence (above diagonal) and standard error of estimates (below diagonal) among mitochondrial haplotypes of A. *mellifera*. Estimates are the weighted averages of divergences calculated from the 48 restriction sites revealed by degenerate and non-degenerate 6-base restriction enzymes. Abbreviations as in Table 8.1.

	lig1	car1	car2	car3	lam1	mel1	mel2	mel3	mel4	int1	int2	int3	int4	scu1	scu2	scu3	scu4	scu5	uni1
lig1	*	0.26	0.51	0.52	2.00	3.16	2.90	2.54	2.80	1.94	2.63	2.54	2.28	3.07	3.07	3.32	3.33	2.80	2.44
car1	0.31	*	0.78	0.26	1.73	3.54	3.28	2.90	3.16	2.28	3.01	2.90	2.63	3.43	3.43	3.69	3.70	3.17	2.83
car2	0.42	0.53	*	0.51	1.94	3.68	3.42	3.06	3.31	2.46	3.17	3.06	2.80	3.57	3.57	3.82	3.82	3.31	3.00
car3	0.44	0.31	0.42	*	1.41	3.81	3.54	3.16	3.42	2.54	3.28	3.17	2.90	3.69	3.69	3.95	3.95	3.43	3.11
lam1	0.96	0.89	0.92	0.78	*	5.11	4.83	4.38	4.65	3.68	4.55	4.38	4.10	4.23	4.23	4.5	4.51	3.95	4.43
mel1	1.03	1.12	1.12	1.17	1.45	*	0.27	0.54	0.26	2.90	3.01	3.55	3.28	4.09	4.09	4.35	4.35	3.82	3.53
mel2	0.98	1.07	1.08	1.12	1.41	0.32	*	0.83	0.54	2.63	2.73	3.28	3.01	3.83	3.83	4.09	4.10	3.55	3.23
mel3	0.88	0.98	0.99	1.03	1.31	0.45	0.58	*	0.26	2.90	3.01	3.55	3.28	4.09	4.09	4.35	4.35	3.82	2.83
mel4	0.94	1.03	1.04	1.08	1.35	0.31	0.45	0.31	*	2.55	2.63	3.17	2.90	3.70	3.70	3.96	3.96	3.43	3.11
int1	0.77	0.85	0.87	0.90	1.11	0.93	0.87	0.93	0.83	*	0.49	0.54	0.27	1.07	1.07	1.32	1.33	0.80	2.16
int2	0.93	1.00	1.01	1.04	1.24	0.97	0.90	0.97	0.87	0.49	*	0.56	0.28	1.10	1.10	1.37	1.37	0.83	2.24
int3	0.90	0.97	0.97	1.01	1.20	1.09	1.03	1.09	0.99	0.47	0.49	*	0.27	0.53	0.53	0.78	0.78	0.26	2.16
int4	0.85	0.93	0.94	0.97	1.17	1.03	0.97	1.03	0.93	0.33	0.35	0.33	*	0.81	0.81	1.07	1.07	0.54	1.87
scu1	1.00	1.09	1.09	1.13	1.25	1.15	1.10	1.15	1.05	0.54	0.56	0.31	0.45	*	0.51	0.25	0.76	0.26	2.73
scu2	1.00	1.09	1.09	1.13	1.25	1.15	1.10	1.15	1.05	0.54	0.56	0.31	0.45	2.19	*	0.76	0.25	0.26	2.73
scu3	1.05	1.13	1.13	1.17	1.30	1.20	1.15	1.20	1.11	0.61	0.64	0.44	0.54	0.31	0.76	*	1.01	0.51	3.01
scu4	1.03	1.12	1.12	1.16	1.30	1.16	1.11	1.16	1.06	0.56	0.58	0.32	0.46	2.66	0.25	0.31	*	0.51	3.02
scu5	0.58	1.01	1.01	1.04	1.14	1.14	1.09	1.14	1.05	0.58	0.60	0.32	0.47	1.56	0.76	0.30	2.19	*	2.44
uni1	0.89	1.00	1.02	1.06	1.38	1.16	1.11	1.00	1.06	0.86	0.90	0.86	0.80	0.97	0.97	1.03	1.00	0.92	*

149

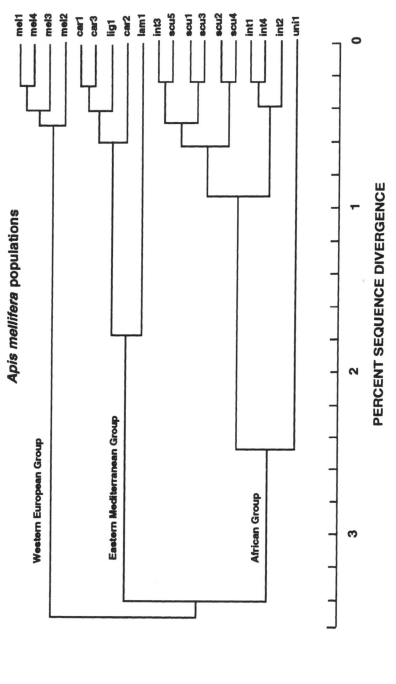

Figure 8.4 Distance phenogram calculated from percent sequence divergence estimates among *A. mellifera* mitochondrial genomes using UPGMA (see Table 8.1 and text for explanation).

different *A. mellifera* mitochondrial haplotypes, as shown in the data matrix in Table 8.1. This suite of enzymes did not reveal differences between *A. m. capensis* and *A. m. scutellata* haplotypes, nor between *A. m. caucasica* and *A. m. carnica* haplotypes. No haplotypes were found that were unique to the subspecies *A. m. iberica*; as discussed below, the mitochondrial haplotypes of *A. m. iberica* were like those of *A. m. mellifera* and *A. m. intermissa*. Thus in subsequent analyses no special mention is made of these three subspecies.

Percent sequence divergences among the 19 haplotypes were calculated from all 48 restriction sites (Table 8.2) A cluster analysis of these data using UPGMA revealed three main groupings of *A. mellifera* haplotypes: an eastern Mediterranean, containing *A. m. ligustica, carnica, caucasica* and *lamarckii*; a western European group containing *A. m. mellifera* and some *A. m. iberica;* and an African group containing *A. m. intermissa, scutellata, capensis, unicolor* and some *iberica* (Figure 8.4).

Two types of mitochondrial DNA were found in Spanish *A. m. iberica*: one similar or identical to that of western European *A. m. mellifera* and one similar or identical to that of north African *A. m. intermissa*. The *A. m. mellifera* mtDNA predominates in samples from the north of Spain, and the north African type predominates in the samples from the south of Spain (Smith et al., 1991). Morphometric and allozyme characters also show a transition from *mellifera*-like in the north to *intermissa*-like in the south (Cornuet and Fresnaye, 1989).

Minor length variation (on the order of 10-20 bp or less) is found throughout the honey bee mitochondrial genome, but two elements of size variation, which appear to be short tandem repeats, are of particular interest. These are a short repeat of approximately 60-80 base pairs in what is probably an A+T rich control region region, and a longer repeat between the COI and COII genes, whose structure was discussed above. The former, which occurs in one, two or three copies has been found only in *A. m. mellifera* (Smith and Brown, 1990). The latter gives rise to five size classes: a small size class, S, found in the eastern Mediterranean group, including *A. m. ligustica, carnica, caucasica* and *lamarckii*; and medium (M), large (L), extra-large (XL) and extra-extra large (XXL) found in the western European and African groups, including *A. m. mellifera, iberica, intermissa, scutellata, capensis* and *unicolor* (Table 8.3; Hall and Smith, 1991).

By combining data on the COI-tRNA[leu] intergenic region and mapped restriction sites, it is possible to present an hypothesis about the phylogenetic relationships of the *A. mellifera* subspecies. All four of the *Apis* species examined have an A+T rich intergenic region; only in *A. cerana*

Table 8.3 Taxonomic distribution of size variation in region betwen COI and COII genes in the mtDNA of *Apis mellifera* subspecies (Hall and Smith, 1991) and *A. cerana* (Cornuet et al., 1991). This size variation is due to an A+T rich intergenic region and a partial duplication of the COI and tRNA^{leu} genes. The small (S) size class has only the A+T rich intergenic sequence; the larger classes -- medium (M), large (L), extra-large (XL) and extra-extra-large (XXL) have one or more tandem duplications of the COI-tRNA^{leu} sequence (Cornuet et al. 1991; Hall and Smith, 1991). In *A. mellifera* the size variation is revealed most clearly in *Bcl* I digests, since two *Bcl* I restriction sites bracket the region of size variation.

Species or Subspecies	Size Class	Size of Bcl I fragment
A. cerana	S	--
A. mellifera		
ligustica,	S	1.05 kb
carnica,	S	
caucasica,	S	
lamarckii	S	
intermissa,	M	1.14 kb
scutellata	M	
unicolor,	M	
mellifera	M	
intermissa,	L	1.32 kb
scutellata,	L	
mellifera	L	
scutellata,	XL	1.56 kb
mellifera	XL	
scutellata	XXL	1.86 kb

and *A. mellifera* (*A. koschevnikovi* was not examined) is there an additional region with sequence homology to COI and tRNA^{leu} (Cornuet et al., 1991). Only in western European and African *A. mellifera* are there tandem repeats of the region with homology to COI and tRNA^{leu}: the M, L, XL and XXL size classes (Table 8.3; Hall and Smith, 1991). Thus, the COI+tRNA^{leu} duplication is a synapomorphy for *A. mellifera* and *A. cerana* (and possibly also *A. koschevnikovi*) and the tandem repeats are a synapomorphy uniting the western European and African *A. mellifera*. The eastern Mediterranean *A. mellifera* subspecies can be considered as the sister taxon to the western European + African subspecies.

Cladistic analysis of the phylogenetic relationships among *A. mellifera* subspecies was carried out using the data matrix in Table 8.1, with the eastern Mediterranean subspecies designated as the outgroup. Twenty-

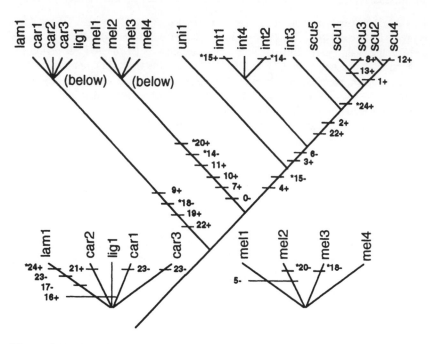

Figure 8.5 Nelson consensus tree for 19 *Apis mellifera* mitochondrial haplotypes. Restriction sites were used as unweighted characters (present or absent). The numbers on the cladogram refer to restriction site characters in Table 8.1; derived character states are indicated by a "+" (present) or "-" (absent). Asterisks indicate characters which show homoplasy, i.e., multiple gains or losses of a character. The terminal haplotypes are: lig1 = *A. m. ligustica*; car1-car3 = *A. m. carnica*; lam1 = *A. m. lamarckii*; mel1-mel4 = *A. m. mellifera*; uni1 = *A. m. unicolor*; int1-int4 = *A. m. intermissa* (from Spanish *A. m. iberica* bees; see text for explanation); scu1-scu5 = *A. m. scutellata*.

eight equally parsimonious trees of 31 steps were found; a Nelson consensus tree (1979), presented in Figure 8.5, had 35 steps and a consistency index of 71. Four characters -- 8, 12, 17 and 21 from Table 8.1 -- were autapomorphies for single terminal taxa; the restriction sites represented by characters 8, 12 and 21 were each present in only one haplotype, and the restriction site represented by character 17 was absent in only one haplotype. If these four non-informative characters are removed from the data matrix the Nelson consensus tree is unchanged and the consistency index is 67 (see Chapter 1, this volume, for discussion of cladistic methodology).

Six characters -- 14, 15, 18, 20, 23, and 24 -- showed homoplasy (Figure 8.5). In characters 14, 18, 20, 23 and 24, homoplasy results when one lineage loses a restriction site, and a terminal taxon in another lineage independently loses the site. For example restriction site 14 is lost only in the *A. m. mellifera*/western European lineage and independently in the haplotype int2 (Figure 8.5). Only character 15 shows independent gain of a restriction site in two taxa. Independent losses of restriction sites are much more likely events than independent gains, since a mutation in any one of the base pairs in a recognition site will cause loss of the recognition site, while independent gains require that two molecules have the exact sequence of base pairs needed for a recognition site.

Biogeographic Implications. The three lineages identified in *A. mellifera* on the basis of mtDNA haplotypes are similar, but not identical, to the lineages identified on the basis of morphometric data. Where the two differ is primarily in the placement of *A. m. intermissa* and *A. m. lamarckii*. Morphometric analysis unites *A. m. intermissa* and other North African subspecies with western European *A. m. mellifera*, with a gradual transition of morphological and allozyme characters from *mellifera*-like values in the north to *intermissa*-like values in the south (e.g., Cornuet and Fresnaye, 1989). The mitochondrial data unites the north African populations (at least *A. m. intermissa*) with the sub-Saharan African subspecies. These three lineages may, as proposed by Ruttner (1968, 1988 and Ruttner et al., 1978), be the result of the colonization of Europe and Africa by three separate lineages of *A. mellifera*. Alternatively, the three lineages could have differentiated in isolation at some time after *A. mellifera* colonized Europe and Africa.

This suggests that the gradual morphological transition across Spain from *A. m. mellifera* to *A. m. intermissa* is the result of secondary contact and gene flow between African and western European lineages in an intraspecific hybrid zone. Similar allozyme transitions are seen in Italy, where the frequency of a malate dehydrogenase allele typical of African populations, $Mdh1^{100}$, decreases in frequency from southern to northern Italy (Badino et al., 1983, 1985), suggesting a similar zone of secondary contact and hybridization between East European and African bees in southern Italy.

The mtDNA of the Madagascar honey bee, *A. m. unicolor*, is related to other African mtDNAs but is clearly different from that of *A. m. scutellata*. It is unlikely that *A. m. unicolor* is evolving so fast that in 2,000 years or less after importation by humans its mtDNA has diverged by 2.4%-3.0% from *A. m. scutellata* mtDNA (typical rates in animal mtDNA are 1% to 2% per million years). Either *A. mellifera* colonized Madagascar at a much earlier date, or human colonists imported another African subspecies whose mtDNA is more similar to that of *A. m.*

unicolor. One obvious candidate is the East African coastal subspecies, *A. m. littorea,* whose mitochondrial genome has not yet been examined.

Future studies of *A. mellifera* mtDNA and biogeography should concentrate on Middle Eastern and African subspecies, which are very poorly known compared to the European subspecies and on the nature of genetic differentiation among *A. mellifera* subspecies. This can be addressed by study of the structure of the apparent hybrid zone between West European and African honey bees in Spain, and by studies of the fitness of intersubspecific hybrids in the wild, especially hybrids between bees of the three major mitochondrial lineages.

Apis cerana and Apis koschevnikovi

Background. As noted above, Maa (1953) placed these bees in their own subgenus, *Apis (Sigmatapis).* Maa recognized 11 species in this group: *A. johni* from Sumatera; *A. lieftincki* from south Sumatera; *A. vechti* from eastern and northern Borneo; *A. nigrocincta* from Sulawesi; *A. javana* from Hainan Island (China), Thailand, peninsular Malaysia, and the Indonesian islands of Sumatera, Java, Karimondjawa, Lombok and Flores; *A. peroni* from the Indonesian island of Timor; *A. samarensis* from the Philippine island of Samar; *A. indica* from India and Sri Lanka; *A. philippina* from the Philippine island of Luzon; *A. cerana* from northern India (Darjeeling), China, and Japan; and *A. koschevnikovi* from "Kamerun". (It is now accepted that the specimens of *A. koschevnikovi* from "Kamerun" examined by Maa were mislabeled, since there are no bees of the *cerana* group in Africa; aside from the misleading locality data, even Maa saw little reason to distinguish these bees from *A. vechti).* Maa's scheme was not widely accepted, and until recently, most authors used *A. cerana* to refer to all of these populations.

Ruttner (1988) reexamined and summarized morphometric information on the eastern cavity-nesting bees, which he considered as one species, *A. cerana.* He grouped the *A. cerana* populations into five subspecies, whose ranges are shown in Figure 8.6: a northern subspecies, *A. cerana cerana* from Afghanistan, Pakistan, north India, China and north Viet Nam; a southern subspecies, *A. c. indica,* from south India, Sri Lanka, Bangladesh, Burma, Malaysia, Thailand, Indonesia and the Philippines; a Japanese subspecies, *A. c. japonica;* and a Himalayan subspecies, *A. c. himalaya.* In essence, Ruttner groups all of Maa's Indonesian, Malaysian and Philippine island species with the southern *A. c. indica,* and splits the Japanese and Himalayan populations away from Maa's *A. cerana.*

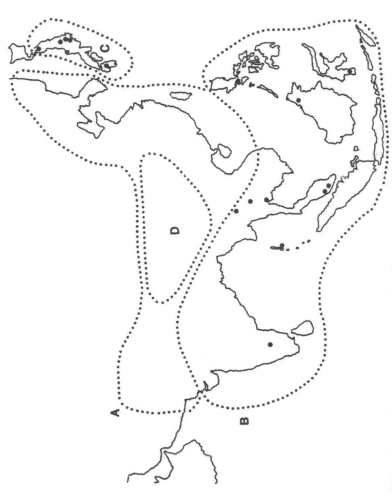

Figure 8.6 Dotted lines indicate the approximate ranges of *Apis cerana* subspecies as recognized by Ruttner (1988). A = *A. cerana cerana*; B = *A. c. indica*; C = *A. c. japonica*; D = *A. c. himalaya*. Large dots show approximate location of collection sites. *Apis koschevnikovi was* collected from Sabah, Borneo at approximately the same locations as *A. cerana*.

Apis cerana populations

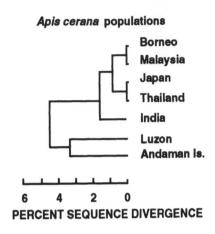

Borneo
Malaysia
Japan
Thailand
India
Luzon
Andaman Is.

6   4   2   0

PERCENT SEQUENCE DIVERGENCE

Figure 8.7 Distance phenogram calculated from percent sequence divergence estimates among *Apis cerana* mitochondrial haplotypes using UPGMA.

Peng et al. (1989) reviewed work by Chinese authors indicating that there are 5 morphologically differentiated populations in China which may represent geographic races similar to those of *A. mellifera*. Statistical analysis of their morphological data by Peng et al. (1989) could confirm clear differences between only two populations or races, but only a small subset of the original data used by the Chinese authors was available to them.

Recently, one of Maa's species, *A. koschevnikovi*, was "rediscovered" on Borneo and Sumatra (Tingek et al., 1988; see Ruttner et al., 1989, for a discussion of the precedence of the name *koschevnikovi* over the name *vechti*). It was found to be sympatric with *A. cerana* (*sensu* Ruttner, 1988) and reproductively isolated both by structure of male endophallus (Tingek et al., 1988) and by time of mating flights (Koeniger et al., 1988).

MtDNA Analysis. For this study, samples of *A. cerana* were obtained from southern India; the Andaman Islands, India; northern and southern Thailand; peninsular Malaysia; northern Borneo (Sabah), Malaysia; the island of Sulawesi, Indonesia; Luzon Island, Philippines; and Japan (Figure 8.6 and Appendix). The four samples from Sulawesi included two color-morphs, black and yellow. The colonies of small, black bees occurred at low elevation in coastal areas and were more defensive, while colonies of the larger yellow bees were found at higher elevation and in more forested areas, and were of a more gentle disposition (G. Otis, pers. comm.; Chapter 2, this volume). Samples of *A. koschevnikovi* were obtained from Sabah, north Borneo (Figure 8.6).

The Japanese samples are the subspecies *A. c. japonica* recognized by Ruttner (1988). All of the other samples belong in the subspecies *A. c. indica* as delimited by Ruttner (1988). Most of the populations recognized as distinct species by Maa (1953) have not been sampled, though the samples from Luzon may correspond to his *A. philippina*, and some or all of the samples from southern Sulawesi may correspond to *A. nigrocincta*.

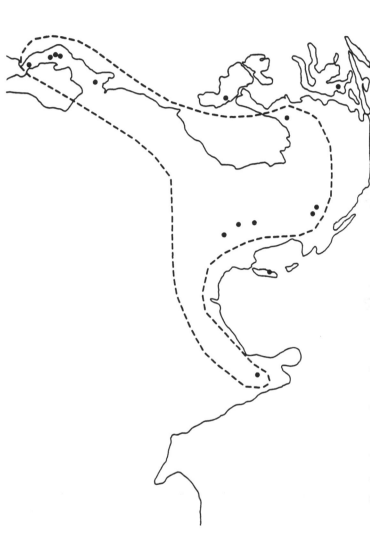

Figure 8.8 *Apis cerana* collection sites superimposed on a map showing the approximate coastline of southeast Asia during the late Pleistocene. Shore lines follow the present-day 200 m bathycline. Dashed lines encircle the mitochondrial DNA haplotypes which differ by 2% or less sequence divergence or less; all of these are located on the Pleistocene mainland. Samples of *A. cerana* taken from islands which remained isolated during the Pleistocene show more divergent mitochondrial haplotypes (see text).

Mitochondrial DNAs prepared from these samples, as well as a sample of *A. mellifera ligustica*, were surveyed with the 6-base restriction enzymes *Bcl* I, *Bgl* II, *Eco*R I, *Hind* III and *Spe* I. Estimates of percent sequence divergence from the resulting restriction fragment data indicated three main lineages of *A. cerana* mtDNA: one including the samples from southern India, Thailand, Malaysia, Borneo and Japan (the "mainland group"); one consisting of the sample from the Andaman Islands; and a third consisting of the samples from Luzon (Figure 8.7).

Sequence divergence within the Asian mainland group ranged from 0% (between Japan and Thailand and between Malaysia and Borneo) to 1.86% (between India and Malaysia/Borneo). Sequence divergence among the three groups was higher: 3.25% between Luzon and the Andamans, 4.06% to 5.62% between Luzon and the mainland samples, and 2.89% to 4.61% between the Andamans and the mainland samples.

A complex situation exists on Sulawesi: two different haplotypes were found, one which is similar, though not identical, to the mainland Asian samples, and a second divergent haplotype. The former haplotype was found in three black bee samples, and the more divergent haplotype was found in the yellow bee sample. Otis suggests that the yellow and black color morphs may represent extreme ecotypes of a single species, sibling species, or colonization of Sulawesi from two divergent source populations of *A. cerana* (Chapter 2, this volume). The four samples examined here are far too few to draw any conclusions; more samples are needed to determine if the two alternate mitochondrial haplotypes are consistently found with the same color morph. If the two color morphs are found to differ in mitochondrial haplotypes, more detailed study of the haplotypes will be needed to determine their phylogenetic affinities.

The mtDNA of *A. koschevnikovi* and *A. mellifera* were both highly divergent from those of the *A. cerana* samples. Pairwise estimates of sequence divergence from the RFLP data gave values of more than 10% between *A. m. ligustica* and each *A. cerana* sample, more than 10% between *A. koschevnikovi* and each *A. cerana* sample, and more than 10% between *A. m. ligustica* and *A. koschevnikovi*. MtDNA sequence divergence among the reproductively isolated species *A. mellifera, koschevnikovi* and *cerana* is much greater than that found among subspecies or populations within *A. mellifera* or *A. cerana*.

Biogeographic Implications. MtDNA data give a picture of *A. cerana* subspecies that differs from that indicated by morphological data. The Japanese population is solidly united with the other populations of the Asian mainland and Borneo that have been sampled, while the island populations from Luzon, the Andamans and probably Sulawesi are quite divergent.

These results make sense in light of the recent geological history of southeast Asia, as shown in Figure 8.8. During the late-middle Pleistocene (approximately 160,000 year ago), sea-level was 160-180 m lower than at present, and at the end of the Pleistocene (about 18,000 years ago) sea-levels were approximately 120 m lower than at present (Heaney, 1986 and references therein). During this time many of the islands of Sunda Shelf (the continental shelf of southeast Asia), including Borneo, Sumatera and Java, were joined to the mainland, collectively forming the region known as Sundaland. Japan and Sri Lanka would also have been joined to the neighboring mainland. The Andamans, Sulawesi, Luzon and other islands were isolated by deep-water channels. Figure 8.8 shows an approximation of the late Pleistocene coastline of southeast Asia.

One possible explanation for the pattern of haplotypes found in *A. cerana* is that populations on the Pleistocene Asian mainland had the opportunity for gene flow and homogenization of haplotypes. Populations on the Andamans, Sulawesi and Luzon remained isolated and continued to diverge from the mainland populations. However, the very fact that islands support relatively small populations may also contribute to their high mtDNA sequence divergence, since novel haplotypes can spread and become fixed by drift much more rapidly in small populations.

Much work remains to be done on *A. cerana* mtDNA. Outstanding issues needing investigation include:

1. Sundaland and the Philippines: the preliminary studies of isolated island populations of *A. cerana* indicate that they (or at least their mitochondrial genomes) are highly divergent from mainland populations. More extensive sampling is needed of the *A. cerana* populations of other southeast Asian islands, particularly the major island groups in the Philippines and the Indonesian Archipelago that have not yet been examined. Many of these islands are believed to have remained isolated throughout the Pleistocene, and may be home to unique *A. cerana* populations. The situation on Sulawesi, discussed above, is just one example of the diversity that may be present.

2. Mainland China: No samples have been examined from the vast region of China and the Himalayas. This region (along with Pakistan and Afghanistan, mentioned below) is home to the northern subspecies *A. c. cerana* and *A. c. himalaya*. In light of

the fact that the mitochondrial genomes of Japanese *A. cerana* are related to those of the southeast Asian mainland, it would appear likely that the mitochondrial genomes of Chinese *A. cerana* will also fall into this group. However this remains to be demonstrated; there may be surprises, particularly in isolated mountain populations.

3. Afghanistan and Pakistan: The *A. cerana* of Afghanistan and Pakistan are the populations in closest geographic proximity to *A. mellifera*. A simplistic but reasonable hypothesis is that the *A. cerana* of Afghanistan and Pakistan are the populations most closely related to *A. mellifera*. Examination of the mtDNA of these populations could shed light on phylogenetic relationships of *A. mellifera* and *A. cerana* suspecies.

Apis dorsata Group

Background. As noted above, Maa (1953) recognized 4 species of giant bees: *Megapis breviligula* from Luzon (and probably elsewhere in the Philippines); *M. binghami* from the Indonesian islands of Sulawesi and Sula; *M. laboriosa* from the Himalayan region; and *M. dorsata*, found over the rest of the range of giant bees. These populations are morphologically rather distinct, and have well defined geographic boundaries. Consequently they have been much more readily accepted as valid taxa, at either the subspecific or specific level (e.g., Sakagami et al., 1980). Ruttner (1988) recognizes each of these as a subspecies of *A. dorsata*; their ranges are shown in Figure 8.9.

MtDNA Analysis. For this study samples of *A. dorsata* were obtained from Pakistan; southern India; the Andaman Islands, India; Thailand; peninsular Malaysia; Sabah, Borneo, Malaysia; and Sulawesi, Indonesia. The samples from Sulawesi are *A. d. binghami*; the rest of the samples are from populations classed as *A. d. dorsata* by Ruttner (1988).

Mitochondrial DNAs prepared from these samples were surveyed with the 6-base restriction enzymes *Afl* II, *Bcl* I, *Bgl* II, *EcoR* I, *Xba* I and *Cla* I; *Hind* III and *Spe* I were tried on most samples but they did not appear to cut these DNAs. Percent sequence divergence estimates made from the resulting restriction fragment data indicated three main lineages of *A. dorsata* mitochondrial haplotypes: one including the samples from Pakistan, the Andamans, Thailand, Malaysia, Borneo (the "mainland group"); a second consisting of the samples from southern India; and a third consisting of the samples from Sulawesi (Figure 8.9 and 8.10).

Figure 8.9 Dotted lines show the approximate ranges of the *A. dorsata* subspecies recognized by Ruttner (1988). A = *A. dorsata dorsata*; B = *A. d. laboriosa*; C = *A. d. breviligula*; D = *A. d. binghami*. Large dots show approximate location of collection sites. The dashed line encircles the "mainland" samples which differed from one another by 1.5% or less sequence divergence. The samples from southern India and Sulawesi were more divergent. (See text for further discussion).

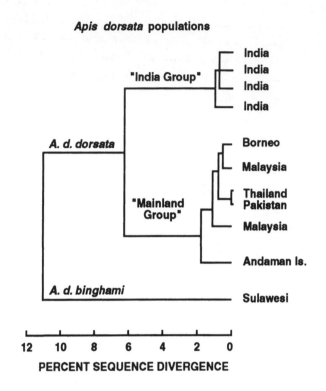

Figure 8.10   Distance phenogram calculated from percent sequence divergence estimates among *A. dorsata* mitochondrial haplotypes using UPGMA.

Sequence divergence within the first group ranged from 0% (between the Pakistani and Thai samples) to 1.99% (between the Andaman Island sample and one Malaysian sample). Sequence divergence among the southern Indian samples ranged from 0% to 1.49%. The two Sulawesi samples were identical. Sequence divergence among the three groups was greater than 5% in all cases: 8.19% to 11.17% between Sulawesi and southern India; 11.72% to 12.23% between Sulawesi and the "mainland group"; and 5.23% to 6.60% between the southern India and the "mainland group".

Biogeographic Implications. The populations of the "mainland group" were quite similar. Unlike *A. cerana*, the mitochondrial genome of the Andaman *A. dorsata* was similar to that of mainland (Pakistani, Thai, Malaysian and Bornean) samples. In addition to their shared mitochondrial restriction fragment patterns, these bees share what appears to be a region of continuous size variation in the mtDNA molecule. In autoradiograms of restriction fragment digests this appears as a broad fuzzy band, consisting of a central dark band and bands of decreasing intensity above it (larger fragments) and below it (smaller fragments). This has not been detected in the samples from Sulawesi or southern India.

Both the Sulawesi and southern Indian samples are highly divergent from the mainland group. In the case of the Sulawesi samples this is not surprising. These bees, *A. d. binghami*, are recognized as morphologically distinct from *A. d. dorsata* and it has been suggested that they are a distinct species (see Chapter 2, this volume, for discussion). And, as was noted in the discussion of *A. cerana*, Sulawesi probably was not joined to the mainland during the Pleistocene, so that both a long period of isolation and small effective population size could contribute to the high level of divergence of this population from the mainland group.

The relatively high level of sequence divergence between southern Indian *A. dorsata* and other samples is much more surprising. If additional study of these mitochondrial haplotypes confirms the high divergence of the southern Indian samples from the other samples, the origin of the divergent mitochondrial haplotypes will be of great interest.

As is true of *A. cerana*, much work remains to be done on the biogeography of *A. dorsata*.

1. The *A. dorsata* haplotypes discussed here must be examined in more detail to confirm the surprisingly high level of divergence between the southern Indian samples and other samples.
2. We have only examined samples belonging to the subspecies (or species) *A. d. dorsata* and *A. d. binghami*. The mitochondrial genomes of Himalayan *A. d. laboriosa* and Philippine *A. d. breviligula* should also be examined.
3. More thorough sampling of the Indian subcontinent is needed. In particular, samples of *A. dorsata* should be collected along transects from southern India northwards to Pakistan in the west and towards Nepal and Bangladesh in the east. This will help determine the true geographic distribution of the mitochondrial haplotypes found in the southern Indian samples. It is possible that this haplotype occurs over a broad region, including most of the regions of India, Bangladesh and Burma which have not been sampled; alternatively, it may be confined to a smaller region of southern India. The form of the contact zone between populations in which the southern Indian haplotype predominates and populations in which the "mainland" haplotype predominates is also of interest.

164

Figure 8.11 Dots show collecting sites for *A. florea* samples (A) and *A. andreniformis* samples (B). Dashed lines indicate the joint ranges of the two dwarf honey bee species.

Apis florea and A. andreniformis

Maa (1953) recognized two dwarf bee species, *A. florea* and *A. andreniformis*. The status of *A. andreniformis* as a separate species from *A. florea* was confirmed by Wongsiri et al. (1990). The ranges of the two species have not been completely explored. *A. florea* has been found from the Persian gulf eastwards to the Thai-Malaysia border, with a disjunct population in Java (Maa, 1953; Ruttner, 1988; Chapter 2, this volume). *A. andreniformis* is found from the Thai-Malaysia border south and east to Java, Borneo and Palawan (Maa, 1953; Chapter 2, this volume). Both species are found throughout Indochina (Chapter 2, this volume). Figure 8.11 shows the approximate ranges of the dwarf bees. For this study samples of *A. florea* were obtained from southern India and Thailand, and samples of *A. andreniformis* were obtained from Malaysia (Borneo and Peninsular; see Appendix). Mitochondrial DNAs prepared from these samples were surveyed with 6-base restriction enzymes *Bcl* I, *Bgl* II, *EcoR* I, *Hind* III and *Spe* I.

There were no differences among three Indian *A. florea* mtDNAs, nor among the three Thai *A. florea*; analysis of shared restriction fragments gave an estimate of 1.99% sequence divergence between Indian and Thai *A. florea*. There were no differences among the three Bornean samples of *A. andreniformis*, but there was some variation among the

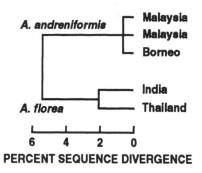

Dwarf honey bee populations

Figure 8.12 Distance phenogram calculated from percent sequence divergence estimates among *A. florea* and *A. andreniformis* mitochondrial haplotypes using UPGMA.

Malaysian *A. andreniformis*; sequence divergence among the three Malaysian samples ranged from 0% to 0.45%. Sequence divergence between the Malaysian and Bornean *A. andreniformis* ranged from 0.45% to 0.91%. Estimates of mtDNA sequence divergence between the two dwarf bee species were higher: 4.61% to 5.96% (Figure 8.12)

As is true of the other Asian *Apis* species, much more sampling is needed for a study of the biogeography of the dwarf bees. Three topics are of particular interest:

1. Investigation of their interactions in areas of sympatry between *A. florea* and *A. andreniformis*.
2. The dwarf bees of the Persian gulf region live in a more arid climate than is typical of dwarf bees. They have unusual nesting habits, sometimes nesting under overhanging rocks without a view of the sky (thought to be necessary for orientation of dances in this species; see Chapter 9, this volume). The status and relationships of this population needs investigation.
3. Dwarf bees are absent from the Philippines, except for the island of Palawan (Ruttner, 1988), which has a particularly interesting history. The channel between Borneo and Palawan is currently 145 m deep; thus Palawan probably has been separated from the Asian mainland since the mid-Pleistocene, approximately 160,000 years ago (Heaney, 1986). The channel between Palawan and Luzon during the mid-Pleistocene would have been very narrow, on the order of 50 km or less (Heaney, 1986). The fact that dwarf bees are found on Palawan but not the other Philippine islands implies that dwarf bees are very poor dispersers across water barriers. Comparison of the mitochondrial genomes of *A. andreniformis* of Borneo and Palawan might give insight into the rates of mito-chondrial evolution in island populations of *Apis*.

Conclusions

We can now return to the five questions posed in the beginning of this study. Each of the species *Apis mellifera, A. cerana* and *A. dorsata* possesses numerous mitochondrial haplotypes with more or less discrete geographic ranges. The samples of *A. koschevnikovi, A. florea* and *A. andreniformis* are too few in number and taken from too small a portion of the species' ranges to assess mitochondrial haplotype diversity in these species.

The mitochondrial haplotypes within *A. mellifera, A. cerana* and *A. dorsata* can be grouped into several major lineages: in *A. mellifera*, an eastern Mediterranean, a western European and an African lineage; in *A. cerana*, a mainland Asian group, a Philippine group, and the Andaman Island population; in *A. dorsata*, the southern Indian, Sulawesi and "main-land + Andaman Islands" group. The general pattern is of distinct mito-chondrial lineages each with a terminal radiation of similar haplotypes that differ from one another by gains or losses of one or a few restriction sites, as illustrated by the distance phenograms in Figures 8.4, 8.7 and 8.10.

Only for *A. mellifera* has this study reached the stage where cladistic methodologies are aplicable to analysis of the relationships among subspecies or haplotypes.

The mitochondrial lineages within each species, and their geographic distributions support some of the traditional *Apis* spubspecies, but not all. In *A. mellifera*, the three mitochondrial lineages are very similar to the three main groupings proposed on the basis of morphological characters. Differences among subspecies within each lineage are more difficult to detect, at least in the eastern Mediterranean samples examined. This may be due to the fairly limited geographic range covered by these samples; more differences among subspecies might become apparent if Middle Eastern populations were included. The western European lineage essentially contains a single subspecies: *A. m. mellifera*. More differentiation has been found among African subspecies, and many more African populations remain to be investigated.

The mitochondrial haplotype data from *A. cerana* do not support the subspecies divisions proposed by Ruttner (1988) very well. Some of the species recognized by Maa (1953) do appear to have distinct mitochondrial haplotypes, but he recognized so many species that it would be surprising if some of them did *not* have unique characteristics. The *A. cerana* lineages indicated by mtDNA data group together all populations that would have been in contact during the mid- to late Pleistocene, when lower sea-levels united much of the Indonesian and Maylaysian archipelagoes with the mainland. Isolated islands such as Luzon have highly distinctive haplotypes.

Mitochondrial haplotype data for *A. dorsata* supports the unique status of *A. d. binghami* from Sulawesi; other populations proposed as subspecies (*A. d. laboriosa* from the Himalayas and *A. d. breviligula* from the Philippines) have not been examined. However, the mtDNA data also suggest that *A. dorsata* from southern India are distinct from the other mainland *A. dorsata*.

The data available thus far are not sufficient to determine conclusively if *A. dorsata, A. cerana* and the dwarf bees have congruent patterns of colonization and differentiation in southeast Asia, but these preliminary data suggest that they do not. In particular, the mitochondrial haplotypes of *A. dorsata* of southern India are distinct from those of *A. dorsata* on the rest of the Asian mainland; there is no such differentiation between the mitochondrial haplotypes of *A. cerana* from India and other parts of the Asian mainland.

Acknowledgments

I wish to thank all of the bee-keepers who kindly allowed me to take samples of their bees, and to the many honey bee researchers who helped me in my work. I am particularly grateful to Dr. Friedrich Ruttner for his encouragment and guidance. Dr. Robert Hagen provided support at all stages of this project: from sample collection to data analysis and criticism of the manuscript. This work was supported by National Science Foundation Grants BSR-8709661, BSR-9010228, and BSR-8918932.

Appendix: Collections

Collections are from nests or hives unless otherwise noted.

Apis mellifera

Collected by D. R. Smith and R. H. Hagen

A. m. mellifera:
1. DENMARK: Laesø Island; 31 July 1987; 10 colonies sampled with help of Mr. Brian Stoklund, Laesø, and Dr. Søren Toft, Zoological Laboratory, University of Aarhus, Aarhus, Denmark; 7 colonies used in this study.
2. NORWAY: Asker, 59°52′N, 10°26′E; 5 August 1987; 8 colonies sampled with help of Mr. Trond Gjessing, Billingstadt, and Dr. Arne Hagen, Bee-Keeping and Bee-Breeding Laboratory, Agricultural University of Norway, Asker; 3 colonies used in this study.
3. SWEDEN: Uppsala, 59°55′N, 17°38′E; 11 August 1987; 3 colonies sampled with help of Dr. Ingmar Fries, Department of Animal Husbandry, Swedish University of Agricultural Science; 3 colonies used in this study.
4. FRANCE: Les Eyzies de Tayac, 44°56′N, 01°02′E, Station Biologique; 28-30 August 1986; 2 colonies sampled with the help of Dr. Roger Darchen, Station Biologique, Les Eyzies; 1 used in this study.
5. FRANCE: Trougemont, near Remiremont, 48°01′N, 06°35′E; September 1988; 3 colonies sampled with the help of Col. Michel LeHeu and Mr. Gabriel Claudel; 1 used in this study.
6. FRANCE: Montfavet, near Avignon, 43°56′N, 04°48′E; September 1988; 5 colonies sampled with the help of Dr. Jean-Marie Cornuet and Mr. Lucien Botella, INRA, Station de Zoologie et d'Apidologie, Montfavet; 2 colonies used in this study.

A. m. carnica
7. AUSTRIA: Graz, 47°05′N, 15°22′E, Steirmarch Imkerschule; 21 July 1987; 5 hives sampled with help of Dr. Hermann Pechhacker, Institut für Bienenkunde, Lunz am See., Austria; 3 colonies used in this study.
8. AUSTRIA: Villach, 46°37′N, 13°51′E; July 1987; 5 hives sampled with help of Mr. Josef Humele, Villach, and H. Pechhacker; 3 colonies used in this study.
9. AUSTRIA: Lunz am See, 47°52′N, 15°02′E; July 1987; 3 colonies sampled with help of H. Pechhacker; 1 colony used in this study.

10. AUSTRIA: Vienna, 48°13′N, 16°22′E; July 1987, 3 colonies sampled from grounds of Institut für Bienenkunde, Lunz am See, Austria, with help of H. Pechhacker; 2 colonies used in this study.
11. YUGOSLAVIA: Medvode, near Kranj, 46°15′N, 14°20′E; July 1987; 3 colonies sampled from grounds of Institut für Bienenkunde, Lunz am See, Austria, with help of H. Pechhacker; 2 colonies used in this study.
12. YUGOSLAVIA: Split, 43°31′N, 16°28′E; July 1987; 3 colonies sampled from grounds of Institut für Bienenkunde, Lunz am See, Austria, with help of H. Pechhacker; 3 colonies used in this study.
13. GERMANY: Hamburg, July 1987; 3 colonies sampled from grounds of Institut für Bienenkunde, Lunz am See, Austria, with help of H. Pechhacker; 1 colony used in this study.

"Nigra" strain: *A. m. mellifera* x *A. m. carnica* hybrids
14. AUSTRIA: Ötztalbanhof, 47°13′N, 10°54′E; July 1987, 9 colonies sampled with the help of Mr. Peter Lettenbichler, Banhof Ötztalbanhof, and H. Pechhacker; 9 colonies used in this study.

A. m. ligustica:
15. ITALY: Sondrio, 46°11′N, 09°52′E, Instituto Sperimentale di Vitifrutticoltura Montana; October 1988, 5 colonies sampled with help of Ms. Guliana. Zuccholi, Morbegno; 1 colony used in study.
16. ITALY: Sondrio, Regoledo di Cosno, 46°02′N, 09°18′E; October 1988, 5 colonies sampled with help of Mr. Enrico Gadola and G. Zuccholi; 5 colonies used in this study.
17. ITALY: Morbegno, 46°08′N, 09°34′E; October 1988, 5 colonies sampled with help of Mr. Italo Buzzeti and Ms. Zuccholi; 5 colonies used in this study.
18. ITALY: Morbegno, 46°08′N, 09°34′E; October 1988, 2 colonies sampled with help of Mr. Paolino Bongio and Ms. Zuccholi; 2 colonies used in this study.
19. ITALY: Milano, 45°28′N, 09°12′E, Milano University; October 1988, 3 colonies sampled with help of Dr. Graziella Serini, Istituto di Entomologia, Universita degli Studi; 3 colonies used in this study.
20. ITALY: Milano, 45°28′N, 09°12′E, Apicultura Moro; October 1988, 5 colonies sampled with help of Mr. Moro; 5 colonies used in this study.
21. ITALY: Bologna, 44°30′N, 11°20′E; October 1988, 5 colonies sampled with help of Dr. Maria A. Vecchi, Istituto Nazionale di Apicultura, Bologna; 4 colonies used in this study.
22. ITALY: Reggio Emilia, 44°42′N, 10°37′E; October 1988, 4 colonies sampled with help of M. A. Vecchi; 1 colony used in this study.

A. m. caucasica:
23. SOVIET UNION: REPUBLIC OF GEORGIA; September 1986; 1 colony sampled from property of Dr. John Kefuss, Toulouse, France.

A. m. iberica
24. SPAIN: Oviedo, 43°21′N, 05°50′W, 10 October 1988; 16 colonies sampled with help of Dr. Anna Quero, Departamento de Biología de Organismos y Sistemas, Universidad de Oviedo, and bee-keepers Mr. José Manuel Menendez, Mr. Julio Alvarez, Mr. José Ramón Fernandez Penedo, Mr. José Isoba Gonzalez, Mr. Pablo Gonzalez Quirós; 16 colonies used in this study.

25. SPAIN: Lugo, 43°00′N, 04°46′W; 11 October 1988; 2 colonies sampled with help of A. Quero; 2 colonies used in this study.
26. SPAIN: Córdoba, 37°53′N, 04°46′W; 15 October 1988; 6 colonies sampled with help of Dr. Francisco Puerta Puerta, Catedra de Biología Aplicada, Seccion de Apicultura, Faculdad de Veterinaria, Universidad de Córdoba, and Dr. Francisco Padilla, Departamento de Ciencias Morfologicas, Faculdad de Veterinaria, Universidad de Córdoba; 6 colonies used in this study.
27. SPAIN: Córdoba, 37°53′N, 04°46′W, Universidad de Córdoba; 15 October 1988; 4 colonies sampled with help of F. Puerta and F. Padilla, 4 colonies used in this study.

Collected by collaborators:

A. m. lamarckii:
28. EGYPT: Giza Governorate, Saqqara, 29°51′N, 31°13′E; 1989; 2 colonies sampled by Steven Goodman, Chicago Field Museum of Natural History, Chicago, Illinois; 2 colonies used in this study.
29. EGYPT: Wadi Gedeed Governorate, Mut, 25°29′N, 28°59′E; 1989; 2 colonies sampled by Steven Goodman; 2 colonies used in this study.

A. m. unicolor:
30. MADAGASCAR: Province Toliara, Fivondronana de Tolagnaro; February 1990; 1 colony sampled by Steven Goodman; 1 colony used in this study.

A. m. scutellata:
31. SOUTH AFRICA: Transvaal, Roodeport, 26°10′S, 27°53′E; 7 colonies sampled by Dr. Robin M. Crewe, Department of Zoology, University of the Witwatersrand, Johannesburg, R.S.A.; 6 colonies used in this study.
32. SOUTH AFRICA: Transvaal, Rust de Winter; 6 colonies sampled by R. M. Crewe; 4 colonies used in this study.

A. m. capensis:
33. SOUTH AFRICA: Cape Province, Stellenbosch, 33°56′S, 18°51′E; 2 colonies sampled by R. M. Crewe; 2 colonies used in this study.
34. SOUTH AFRICA: Cape Province, Klapmuts, 2 colonies sampled by R. M. Crewe; 2 colonies used in this study.

Apis koschevnikovi

1. MALAYSIA: BORNEO: Sabah, Tenom Agricultural Research Station, 5°10′N, 115°57′E; 16 June 1989; collected by Dr. Gard Otis, Dept. Environmental Biology, University of Guelph. Guelph, Ontario CANADA (=samples GWO95 and GWO96).

Apis cerana

A. c. indica (*sensu* Ruttner, 1988);
1. INDIA: Karnataka State, Bangalore, 12°58′N, 77°35′E; May 1989; 1 colony sampled by Dr. Fred C. Dyer, Dept. Zoology, Michigan State University, East Lansing, MI USA.
2. INDIA: Karnataka State, Bangalore, 12°58′N, 77°35′E; August 1990; 11 colonies sampled by D. R. Smith and F. C. Dyer; 2 used in this study.

3. INDIA: ANDAMAN ISLANDS, Port Blair, 11°40′N, 92°44E; August 1990, sampled from flowers by G. Otis.

4. MALAYSIA: Selangor, Serdang, Universiti Pertanian Malaysia, 3°0′N, 101°41′E; 12 March 1989; collected by G. Otis (=samples GWO18, GWO20).

5. MALAYSIA: Pahang, Fraser's Hill, 3°43′N, 101°44′E; 1 April 1989; collected by G. Otis (=sample GWO22).

6. MALAYSIA: BORNEO, Sabah, Tenom Agricultural Research Station, 5°10′N, 115°57′E; 5 May 1989; collected by G. Otis (= samples GWO103, GWO105, GWO106, GWO107).

7. THAILAND: Pitsanulok, 16°47′N, 100°13′E; 8 April 1989; collected by G. Otis (= sample GWO28).

8. THAILAND: Chang Mai region; February 1988; 5 colonies sampled by Hermann Pechhacker, Institut für Bienenkunde, Lunz am See, Austria, Dr. Aht Boonatee, and Niran Juntawong; 2 samples used in this study.

9. THAILAND: Bangkok region; February 1988; 3 colonies sampled by H. Pechhacker, A. Boonatee and N. Juntawong; 1 colony used in this study (S15).

10. PHILIPPINES: LUZON: Benguet Prov., ~15 km west of Baguio, 16°29′N, 120°31′E; 27 May 1989; collected by G. Otis (=sample GWO62).

11. PHILIPPINES: LUZON: Zambales Prov., Santa Cruz, 15°52′N, 119°53′E; 27 May 1989; collected by G. Otis (=sample GWO63).

12. PHILIPPINES: LUZON: Laguna Prov., Los Baños; 28 May 1989; collected by G. Otis (=samples GWO64-GWO69).

13. INDONESIA: SULAWESI: Jeneponto, 5°42′S, 119°41′E; 8 June 1989; collected by G. Otis (=sample GWO91).

14. INDONESIA: SULAWESI: Ujung Pandang 5°6′S, 119°30′E; 2 June 1989; collected by G. Otis (=sample GWO76).

15. INDONESIA: SULAWESI: Tabo Tabo Forestry Center 4°50′S, 119°38′E; 3 June 1989; collected by G. Otis (=sample GWO77).

16. INDONESIA: SULAWESI: W. Sinjai district; 6 June 1989; collected by G. Otis (=sample GWO83).

17. INDONESIA: SULAWESI: 12 km N. of Bulukumba, 05°35′S, 120°13′E; 8 June 1989; collected by G. Otis (=sample GWO89).

A. cerana japonica (Ruttner, 1988)

18. JAPAN: Honshu Island, Tokyo Prefecture, Tokyo; September 1990; 3 colonies sampled with help of Drs. Masami Sasaki and Tadaharu Yoshida, Institute of Honeybee Science, Tamagawa University, Tokyo; 1 colony used in this study.

19. JAPAN: Honshu Island, Kanagawa Prefecture, Kanagawa; September 1990; 1 colony sampled with help of M. Sasaki and T. Yoshida; 1 colony used in this study.

20. JAPAN: Kyushu Island, Kumamoto Prefecture; September 1990; 2 colonies sampled withh help of M. Sasaki and T. Yoshida; 1 colony used in this study.

21. JAPAN: Honshu Island, Iwate Prefecture, Morioka, 39°43′N, 141°08′E; September 1990; 10 colonies sampled with help of Mr. S. Fujiwara; 1 colony used in this study.

22. JAPAN: Honshu Island, Tochigi Prefecture, Nasu-Shiobara, near Nasu-Yumoto, 37°06′N, 140°00′E; September 1990; 2 colonies sampled with help of Mr. D. Shimotori; 1 colony used in this study.

Apis dorsata

A. dorsata dorsata (*sensu* Ruttner, 1988)
1. INDIA: Karnataka State, Bangalore, 12°58′N, 77°35′E; May 1989; 1 colony sampled by F. C. Dyer.
2. INDIA: Karnataka State, Bangalore, 12°58′N, 77°35′E; August 1990; 7 colonies sampled by D. R. Smith and F. C. Dyer; 5 used in this study.
3. INDIA: ANDAMAN ISLANDS, Port Blair, 11°40′N, 92°44′E; 2 August 1990; collected from flowers by G. Otis.
4. PAKISTAN: Islamabad, 33°40′N, 73°08′E, Ambassador Hotel; 1990; 1 colony collected by S. Goodman.
5. MALAYSIA: Selangor, Serdang, 3°0′N, 101°41′E; 8 February 1989; collected by G. Otis (=sample GWO3).
6. MALAYSIA: Kedah, nr. Sik, 5°50′N, 100°57′E; 16 February 1989; collected by G. Otis (=samples GWO4-GWO7).
7. MALAYSIA: Perak, Sitiawan, 04°11′N, 100°42′E; 13 March 1989; collected by G. Otis, (=sample GWO19).
8. MALAYSIA: BORNEO: Sabah, Tenom Agricultural Research Station, 5°10′N, 115°57′E; 16 June 1989; collected by G. Otis (=samples GWO97-99).
9. THAILAND: Bangkok, 13°44′N, 100°30′E; 7 April 1989; collected by G. Otis, (=sample GWO23).
10. THAILAND: Chang Mai region; February 1988; 1 colony sampled by H. Pechhacker, A. Boonatee and N. Juntawong.

A. dorsata binghami
11. INDONESIA: SULAWESI: Tabo Tabo Forestry Center 4°50′S, 119°38′E; 3-4 June 1989; collected by G. Otis (=sample GWO80,81).

Apis florea

1. INDIA, Bangalore, 12°58′N, 77°35′E; May 1989; 1 colony sampled by F. C. Dyer.
2. INDIA, Bangalore, 12°58′N, 77°35′E; August 1990; 6 colonies sampled by D. R. Smith and F. C. Dyer; 5 used in this study.
3. THAILAND: Chang Mai region; February 1988; 5 colonies sampled by H. Pechhacker, A. Boonatee and N. Juntawong; 2 samples used in this study.
4. THAILAND: Nakhom Sawan, 300 km N. Bangkok, 15° 42′N, 100° 7′E; 8 April 1989; collected by G. Otis (=sample GWO29).
5. THAILAND: Bangkok; 28 February and 7 April 1989; collected by G. Otis (=samples GWO8,9,10,27).

Apis andreniformis

1. MAYLAYSIA: Johor, 1°55′N, 102°46′E; 27 & 30 January 1989 and 2 March 1989; collected by G. Otis (=samples GWO1,2,13,15,16).
2. MAYLAYSIA: BORNEO: Sabah, Tenom Agricultural Research Station, 5°10′N, 115°57′E; 17-18 June 1989; collected by G. Otis (=samples GWO100,101,102,113).

Literature Cited

Adam, O. B. E., Brother. 1983. *In search of the best strains of honey bee.* Northern Bee Books.

Anderson, S., A. T. Bankier, B. G. Barrell, M. H. L. De Bruijn, A. R. Coulson, J. Drouin, I. C. Eperon, D. P. Nierlich, B. A. Roe, F. Sanger, P. H. Schreier, A. J. H. Smith, R. Staden and I. G. Young. 1981. Sequence and organization of the human mitochondrial genome. *Nature* 290:457-465.

Anderson, S., M. H. L. De Bruijn, A. R. Coulson, I. C. Eperon, F. Sanger and I. G. Young. 1982. The complete sequence of bovine mitochondrial DNA: conserved features of the mammalian mitochondrial genome. *J. Mol. Biol.* 156:683-717.

Anon. 1988. The honeybee of Madagascar. *Bee World* 69:164-166.

Badino, G., G. Celebrano and A. Manino. 1983. Population structure and *Mdh-1* locus variation in *Apis mellifera ligustica. J. Hered.* 74:443-446.

Badino, G., G. Celebrano, A. Manino and S. Longo. 1985. Enzyme polymorphism in the Sicilian honeybee. *Experientia* 41:752-754.

Bibb, M. J., R. A. Van Etten, C. T. Wright, M. W. Walberg and D. A. Clayton. 1981. Sequence and gene organization of mouse mitochondrial DNA. *Cell* 26:167-180.

Boyce, T. M., M. E. Zwick and C. F. Aquadro. 1989. Mitochondrial DNA in the bark weevils: size, structure and heteroplasmy. *Genetics* 123:825-836.

Brückner, D. 1974. Reduction of biochemical polymorphisms in honeybees (*Apis mellifica*). *Experientia* 30:618-619.

Brown, W. M. 1985. The mitochondrial genome of animals. in *Molecular Evolutionary Genetics.*, R. J. MacIntyre, ed. Plenum Press: New York, NY. Pp. 95-130.

Brown, W. M., E. M. Prager, A. Wang, and A. C. Wilson. 1982. Mitochondrial DNA sequences of primates: tempo and mode of evolution. *J. Mol. Evol.* 18:225-239.

Cantatore, P., M. Roberti, G. Rainaldi, M. N. Gadaleta and C. Saccone. 1989. The complete nucleotide sequence, gene organization, and genetic code of the mitochondrial genome of *Paracentrotus lividus. J. Biol. Chem.* 264:10965-10975.

Clary, D. O. and D. R. Wolstenholme. 1985. The mitochondrial DNA molecule of *Drosophila yakuba*: nucleotide sequence, gene organization, and genetic code. *J. Mol. Evol.* 22:252-271.

Cockburn, A. F., S. E. Mitchell and J. A. Seawright. 1990. Cloning of the mitochondrial genome of *Anopheles quadrimaculatus. Arch. Insect Biochem. Physiol.* 14:31-36.

Cornuet, J.-M. and J. Fresnaye. 1989. Étude biométrique de colonies d'abeilles d'Espagne et du Portugal. *Apidology* 20:93-101.

Cornuet, J.-M., L. Garnery and M. Solignac. 1991. Putative origin and function of the intergenic region between COI and COII of *Apis mellifera* L. mitochondrial DNA. *Genetics* 1128:393-403.

Crozier, R. H., Y. C. Crozier and A. G. Mackinlay. 1989. The CO-I and CO-II region of honeybee mitochondrial DNA: evidence for variation in insect mitochondrial evolutionary rates. *Mol. Biol. Evol.* 6:399-411.

Crozier, Y. C., S. Koulianos and R. H. Crozier. 1991. An improved test for Africanized honeybee mitochondrial DNA. *Experientia*, submitted.

de Bruijn, M. H. L. 1983. *Drosophila melanogaster* mitochondrial DNA, a novel organization and genetic code. *Nature* 304:234-241.

Desjardins, P. and R. Morais. 1990. Sequence and gene organization of the chicken mitochondrial genome A novel gene order in higher vertebrates. *J. Mol. Biol.* 212:599--634.

Dubin, D. T., C. C. HsuChen and L. E. Tillotson. 1986. Mosquito mitochondrial transfer RNAs for valine, glycine, and glutamate: RNA and gene sequences and vicinal genome organization. *Curr. Genet.* 10:701-707.

Farris, J. S. 1988. Hennig86 Reference, Version 1.5. Port Jefferson Station, New York.

Garesse, R. 1988. *Drosophila melanogaster* mitochondrial DNA: gene organization and evolutionary considerations. *Genetics* 118:649-663.

Garnery, L., D. Vautrin, J.-M. Cornuet and M. Solignac. 1991. Phylogenetic relationships in the genus *Apis* inferred from mitochondrial DNA sequence data. *Apidologie*, in press.

Gyllensten, U., D. Wharton, and A. C. Wilson. 1985. Maternal inheritance of mitochondrial DNA during backcrossing of two species of mice. *J. Hered.* 76:321-324.

Hall, H. G. and K. Muralidharan. 1989. Evidence from mitochondrial DNA that African honey bees spread as continuous maternal lineages. *Nature* 339:211-213.

Hall, H. G. and D. R. Smith. 1991. Distinguishing African and European honeybee matrilines using amplified mitochondrial DNA. *Proc. Natl. Acad. Sci. USA* 88:4548-4552.

Haucke, H.-R. and G. Gellison. 1988. Different mitochondrial gene orders among insects: exchanged tRNA gene positions in the COII/COIII region between orthopteran and dipteran species. *Curr. Genet.* 14:471-476.

Heaney, L. R. 1986. Biogeography of mammals in SE Asia: estimates of rates of colonization, extinction and speciation. *Biol. J. Linn. Soc.* 28:127-165.

Hoeh, W. R., K. H. Blakely and W. M. Brown. 1991. Heteroplasmy suggests limited biparental inheritance of *Mytilus* mitochondrial DNA. *Science* 251:1488-1490.

HsuChen, C. C. and D. T. Dubin. 1984. A cluster of four transfer RNA genes in mosquito mitochondrial DNA. *Biochem. Int.* 8:385-391.

HsuChen, C. C., R. M. Kotin, and D. T. Dubin. 1984. Sequences of the coding and flanking regions of the large ribosomal subunit RNA gene of mosquito mitochondria. *Nucleic Acids Res.* 12:7771-7785.

Jacobs, H., D. Elliot, V. B. Math and A. Farquharson. 1988. Nucleotide sequence and gene organization of sea urchin mitochondrial DNA. *J. Mol. Biol.* 201:185-217.

Koeniger, N., G. Koeniger, S. Tingek, M. Mardan, and T. E. Rinderer. 1988. Reproductive isolation by different time of drone flight between *Apis cerana* (Fabricius 1793) and *Apis koschevnikovi* (Buttel-Reepen 1906). *Apidologie* 19:103-106.

Kondo, R., Y. Satta, E. T. Matsuura, H. Ishiwa, N. Takahata and S. Chigusa. 1990. Incomplete maternal transmission of mitochondrial DNA in *Drosophila*. *Genetics* 126:657-663.

Lansman, R. A., J. C. Avise, and M. D. Huettel. 1983. Critical experimental test of the possibility of "paternal leakage" of mitochondrial DNA. *Proc. Natl. Acad. Sci. USA* 80:1969-1971.

Maa, T.-C. 1953. An inquiry into the systematics of the tribus Apidini or honeybees (Hym.). *Treubia* 21:525-640.

McCracken, A., I. Uhlenbusch and G. Gellissen. 1987. Structure of the cloned *Locusta migratoria* mitochondrial genome: restriction mapping and sequence of its ND-1 (URF-1) gene. *Curr. Genet.* 11:625-630.

Meixner, M., F. Ruttner, N. Koeniger and G. Koeniger. 1989. The mountain bees of the Kilimanjaro region and their relation to neighboring bee populations. *Apidologie* 20:165-174.

Mestriner, M. A. 1969. Biochemical polymorphisms in bees (*Apis mellifera ligustica*). *Nature* 223:188-189.

Mestriner, M. A. and E. P. B. Contel. 1972. The P-3 and Est loci in the honeybee, *Apis mellifera*. *Genetics* 72:733-738.

Meusel, M. and R. F. A. Moritz. 1990. Transfer of paternal mitochondrial DNA in fertilization of honeybee (*Apis mellifera* L.) eggs. In *Social Insects and the Environment.*, G.K. Veeresh, B. Mallik, C.A. Viraktamath, eds. Oxford and IBH Publ. Co.: New Delhi. P. 135.

Moritz, C., T. E. Dowling and W. M. Brown. 1987. Evolution of animal mitochondrial DNA: Relevance for population biology and systematics. *Ann. Rev. Ecol. Syst.* 18:269-292.

Moritz, R. F. A., C. F. Hawkins, R. H. Crozier and A. G. Mackinlay. 1986. A mitochondrial DNA polymorphism in honeybees (*Apis mellifera* L.). *Experientia* 42:322-324.

Nei, M. 1987. *Molecular Evolutionary Genetics.* Columbia University Press: New York, NY.

Nei, M. and F. Tajima, F. 1983. Maximum likelihood estimation of the number of nucleotide substitutions from restriction sites data. *Genetics* 105:207-217.

Nelson, G. 1979. Cladistic analysis and synthesis: Principles and definitions, with a historical note on Adanson's "Familles des plantes" (1763-1764). *Syst. Zool.* 28:1-21.

Owen, H. G. 1983. *Atlas of continental displacement 200 million years to the present.* Cambridge University Press: Cambridge.

Peng, Y. S., M. E. Nasr and S. J. Locke. 1989. Geographical races of *Apis cerana* Fabricius in China and their distribution. Review of recent Chinese publications and a preliminary statistical analysis. *Apidology* 20:9-20.

Rinderer, T. E., N. Koeniger, S. Tingek, M. Mardan and G. Koeniger. 1989. A morphological comparison of the cavity dwelling honeybees of Borneo *Apis koschevnikovi* (Buttel-Reepen, 1906) and *A. cerana* (Fabricius, 1793). *Apidologie* 20:405-411.

Roe, B. A., D. P. Ma, R. K. Wilson and J. F. H. Wong. 1985. The complete nucleotide sequence of the *Xenopus laevis* mitochondrial genome. *J. Biol. Chem.* 260:9759-9774.

Ruttner, F. 1968. Les races d'abeilles. in *Traité de biologie de l'abeille, Vol. I, Biologie et physiologie générales.*, R. Chauvin, ed. Mason et Cie.: Paris

------. 1980. *Apis mellifera adami* (n. ssp.), die Kretische Biene. *Apidolgie* 11:385-400.

------. 1988. *Biogeography and taxonomy of honey bees.* Springer-Verlag: Berlin.

Ruttner, F., D. Kauhausen and N. Koeniger. 1989. Position of the Red honey bee, *Apis koschevnikovi* (Buttel-Reepen 1906), within the genus *Apis. Apidologie* 20:395-404.

Ruttner, F., D. Pourasghar, and D. Kauhausen. 1985. Die honigbienen des Iran 2. *Apis mellifera meda* Skorikow, die Persische Biene. *Apidologie* 16:241-264.

Ruttner, F., L. Tassencourt and J. Louveaux. 1978. Biometrical-statistical analysis of the geographic variability of *A. mellifera* L. *Apidologie* 9:363-381.

Saiki, R. K., S. Scharf, F. Faloona, K. B. Mullis, G. T. Horn, H. A. Erlich and N. Arnheim. 1985. Enzymatic amplification of β-globin genomic sequences and restriction site analysis for diagnosis of sickle cell anemia. *Science* 230:1350-1354.

Sakagami, S., T. Matsumura and K. Ito. 1980. *Apis laboriosa* in Himalaya, the little known world largest honeybee (Hymenoptera, Apidae). *Insecta Matsumurana* New Series 19: 47-77.

Satta, Y., N. Toyohara, C. Ohtaka, Y. Tatsuno, T. K. Watanabe, E. T. Matsuura, S. I. Chigusa and N. Takahata. 1988. Dubious maternal inheritance of mitochondrial DNA in *D. simulans* and evolution of *D. mauritiana. Genet. Res.* 52:1-6.

Sheppard, W. S., T. E. Rinderer, J. A. Mazzoli, J. A. Stelzer and H. Shimanuki. 1991. Gene flow between African- and European-derived honey bee populations in Argentina. *Nature* 349:782-784.

Simon, C. 1991. Molecular systematics at the species boundary: exploiting conserved and variable regions of the mitochondrial genome of animals via direct sequencing from amplified DNA. in *Molecular Taxonomy.*, G. M. Hewitt, A. W. B. Johnston and J. P. W. Young, eds. Nato Advanced Studies Institute, Springer Verlag: Berlin. In press.

Simon, C., A. Francke and A. Martin. 1991. The polymerase chain reaction: DNA extraction and amplification. in _Molecular Taxonomy._, G. M. Hewitt, A. W. B. Johnston and J. P. W. Young, eds. Nato Advanced Studies Institute, Springer Verlag:Berlin. In press.

Smith, D. R. 1988. Mitochondrial DNA polymorphisms in five Old World subspecies of honey bees and in New World hybrids., in _Africanized honey bees and bee mites._ G. R. Needham, R. E. Page, Jr., M. Delfinado-Baker, and C. E. Bowman, eds. 1988. Ellis Horwood, Ltd.: Chichester. Pp. 303-312.

------. 1991. African bees in the Americas: insights from biogeography and genetics. _TREE_ 6:17-21.

Smith, D. R. and W. M. Brown. 1988. Polymorphisms in mitochondrial DNA of European and Africanized honeybees (_Apis mellifera_). _Experientia_ 44:257-260.

------. 1990. Restriction endonuclease cleavage site and length polymorphisms in mitochondrial DNA of _Apis mellifera mellifera_ and _A. m. carnica_ (Hymenoptera: Apidae). _Ann. Entomol. Soc. Am._ 83:81-88.

------. 1991. A mitochondrial DNA restriction enzyme cleavage site map for the scorpion _Hadrurus arizonensis_ (Iuridae). _J. Arachnol._ in press.

Smith, D. R., M. F. Palopoli, B. R. Taylor, L. Garnery, J.-M. Cornuet, M. Solignac and W. M. Brown. 1991. Geographical overlap of two mitochondrial genomes in Spanish honey bees (_Apis mellifera iberica_). _J. Hered._ 82:96-100.

Smith, D. R., O. R. Taylor and W. M. Brown. 1989. Neotropical African bees have African mitochondrial DNA. _Nature_ 339:213-215.

Sneath, P. H. A. and R. R. Sokal. 1973. _Numerical taxonomy._ W. H. Freeman and Co.: San Francisco.

Snyder, M., A. R. Fraser, J. LaRoche, K. E. Gartner-Kepkay and E. Zouros. E. 1987. Atypical mitochondrial DNA from the deep-sea scallop _Placopecten magellanicus_. _Proc. Natl. Acad. Sci. USA_ 84:7595-7599.

Tingek, S., M. Mardan, T. E. Rinderer, N. Koeniger and G. Koeniger. 1988. Rediscovery of _Apis vechti_ (Maa, 1953): the Saban honey bee. _Apidologie_ 19:97-102.

Uhlenbusch, I., A. McCracken and G. Gellissen. 1987. The gene for the large (16S) ribosomal RNA from the _Locusta migratoria_ mitochondrial genome. _Curr. Genet._ 11:631-638.

Vlasak, I., S. Burgschwaiger and G. Kreil. 1987. Nucleotide sequence of the large ribosomal RNA of honeybee mitochondria. _Nucleic Acids Res._ 15:2388.

Wolstenholme, D. R., J. L. MacFarlane, R. Okimoto, D. O. Clary and J. A. Wahleithner. 1987. Bizarre tRNAs inferred from DNA sequences of mitochondrial genomes of nematode worms. _Proc. Natl. Acad. Sci. USA_ 84:1324-1328.

Wongsiri, S., K. Limbipichai, P. Tangkanasing, M. Mardan, T. Rinderer, H. A. Sylvester, G. Koeniger and G. Otis. 1990. Evidence of reproductive isolation confirms that _Apis andreniformis_ (Smith, 1858) is a separate species from symaptric _Apis florea_ (Fabricius, 1787). _Apidologie_ 21:47-52.

Wu, Y. and B. Kuang. 1987. Two species of small honeybee - a study of the genus _Micrapis_. _Bee World_ 68:153-155.

9

Comparative Studies of Dance Communication: Analysis of Phylogeny and Function

Fred C. Dyer

Introduction

The dance language of honey bees, by which foragers communicate the location of food to their nest mates, is one of the most interesting and best studied aspects of their biology. In precision and flexibility, it stands apart from the communication systems of all other insects, raising compelling questions about how it evolved. Since Karl von Frisch (1967, review) first decoded the dance *of Apis mellifera* in the 1940s, most studies of it have concerned either the mechanisms by which bees obtain the spatial information that they express in dances (review by Dyer and Seeley, 1989a), or the social processes by which dances are integrated into a colony-level foraging strategy (reviews by Seeley, 1985, 1989). Until recently, however, far less effort has been focused on the evolutionary questions of how this remarkable behavior was assembled from simpler behavioral elements, and how it was subsequently shaped by natural selection. My goal in this chapter is to review recent investigations of these evolutionary questions; I focus especially on studies that have drawn their insights from the diversity in dance communication among different *Apis* species.

Honey bees are ideally suited for comparative behavioral studies. The *Apis* species and subspecies exhibit a variety of distinct behavioral differences, presumably corresponding to distinct evolutionary transitions; studies of character evolution require such diversity as a starting point.

177

Furthermore, honey bees form a monophyletic group (Ruttner, 1988), potentially simplifying studies of both phylogeny and adaptation. Finally, the decades of research on the European races of *Apis mellifera* have generated both a useful basline for comparison and an armamentarium of powerful experimental techniques.

The comparative studies I shall discuss all build on Lindauer's (1956) pioneering work on the three other species of *Apis* known at that time: the Asian hive bee, *A. cerana*, the giant honey bee (or rock bee), *A. dorsata*, and the dwarf honey bee, *A. florea*. All three are distributed widely across southern Asia; Lindauer studied them in Sri Lanka. Like *A. mellifera*, all are eusocial, reproduce by colony fission, build wax combs, and have dance languages. They differ from each other, however, in body size, in nesting behavior, and in certain properties of their dances. On the basis of correlated interspecific differences in dancing behavior and nest architecture, Lindauer put forth a now classic hypothesis about the phylogenetic origin and modification of the code by which honey bees communicate flight direction. Also, in a separate set of observations, he determined the code by which each of the species in Sri Lanka communicates flight distance, and provided evidence that interspecific differences in the code were the adaptive outcome of selective pressures experienced differently by the three species.

The bulk of this chapter consists of three case studies. The first two directly extend the investigations begun by Lindauer on the phylogeny of the direction code (Dyer, 1984, 1985a,b, 1987; Dyer and Seeley, 1989a) and on the adaptive design of the distance code (Dyer and Seeley, 1991a). The third concerns an additional line of work (Towne, 1985) on the sounds produced by dancing bees. These case studies clarify our understanding of particular aspects of the evolution of the dance language. In addition, they illustrate the following general issues. First, hypotheses about the phylogeny and function of behavioral characters may be greatly hampered by the difficulty of describing the behavior, partitioning it into more fundamental characters, or homologizing these characters with traits exhibited by other taxa. Second, a phylogenetic hypothessis may be critical for interpreting the adaptive sigificance of traits. To the extent that the studies I describe come to grips with these important issues, I believe that they strengthen the value of honey bees as model organisms for the study of behavioral evolution.

Before I continue, I shall briefly explain my decision to discuss only the four *Apis* species which were the focus of Lindauer's study. In recent years at least three new species have been recognized: *A. andreniformis*, *A. koschevnikovi*, and *A. laboriosa* (reviewed in Chapter 2, this volume).

In worker morphology and nesting behavior, each resembles one of the Asian species that Lindauer studied (*florea, cerana,* and *dorsata,* respectively) closely enough to have been previously considered a subspecies (Ruttner, 1988). I shall assume, pending further study of these little-investigated taxa, that the resemblances extend to the principal features of the dance language, hence that the traditional classification based on four species (the three Asian species plus *A. mellifera*) captures all of the relevant behavioral diversity.

The Dance Language and Its Components

The behavior of *Apis mellifera* illustrates several basic features of honey bee dance communication. Forager bees perform dances in the nest after returning from sources of food, water, or propolis (von Frisch, 1967, Seeley, 1985, Winston, 1987). If the source is nearby, the forager does a "round dance," simply turning in circles in the nest, regularly alternating the direction of circling. (Although subtle directional information is potentially available in round dances [Kirchner et al., 1988], bees may not use it: von Frisch, [1967, pp 150-151] observed that round dances cause recruits to search in all directions within a few tens of meters from the nest.) If the food source is more distant, a bee communicates both the distance and direction of flight by performing a "waggle dance," a series of movements in which she waggles her body from side to side and buzzes her wings while running in a particular direction on the comb (Figure 9.1). The distance signal is apparently given in the duration in time of each repeated waggling run (von Frisch, 1967); this and several other correlated features of the dance increase monotonically with flight distance according to a population-specific function. The direction of flight, which the bee has measured by reference to the sun and sun-linked polarization patterns in blue sky (reviews by Wehner, 1984; Dyer and Seeley, 1989b), is expressed in the orientation of the runs relative to environmental features detected during the dance. In *A. mellifera*'s dances, which are normally performed in the darkness of the nest, gravity usually provides the directional reference: the dance angle relative to the upward direction on the comb indicates the flight direction relative to the solar azimuth (solar flight angle). When dancers have a view of celestial cues, however, for example when dancing on the surface of a swarm cluster, they ignore gravity and orient to the sun itself (von Frisch, 1967).

To analyze the evolution of any complex behavior such as the dance language, a necessary first step is to partition the stream of behavior into components assumed to represent more fundamental phenotypic units (Lauder, 1986). This provides a set of characters whose evolutionary changes can be analyzed, analogous to the sets of characters available to morphologists. To the extent that these characters are homologous with ancestral behavioral elements which may also be expressed (in other behavioral contexts) in living relatives of the species under study, such an analysis may reveal the historical origins of the trait in question. Furthermore, hypotheses about the adaptive modification of a complex trait often require the assumption that its components can be shaped independently (Krebs and McCleery, 1984).

Though decisions about the initial partitioning of a behavioral trait are often ambiguous (Dawkins, 1983), even with knowledge of the underlying neuromuscular mechanisms (Lauder, 1986), two sources of evidence may validate them. First, different components may correspond to clearly separable functions composing the overall function of the behavior. Second, components may be expressed independently of one another within a species, or may vary independently among different species.

The traditional way of partitioning the dance language (von Frisch, 1967), which I shall follow and extend in this paper, is implicitly based on both of these sources of evidence. For example, one major division separates the communication of distance (waggling run duration) from the communication of direction (waggling run orientation), which are quite plausibly assumed to be under separate neural control and based on separate sources of information acquired during the flight. The distance code is usually treated as a unitary trait corresponding to a putative set of neural mechanisms which translates flight distance into the distance signal (Lindauer, 1985). The direction code is partitioned further into the mechanisms by which dancers and dance followers measure their orientation relative to celestial cues and the mechanisms by which

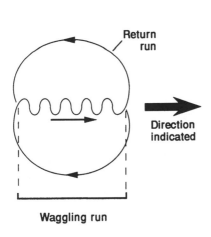

Figure 9.1 Sketch of the waggle dance. The dancer moves repeatedly through a roughly figure-eight pattern; both flight direction and flight distance are communicated in the waggling run, represented by the wavy line. Distance is given in its duration. Direction in its orientation relative to an environmental reference. As explained in the text, *Apis mellifera* orients relative to either celestial cues or the direction of gravity.

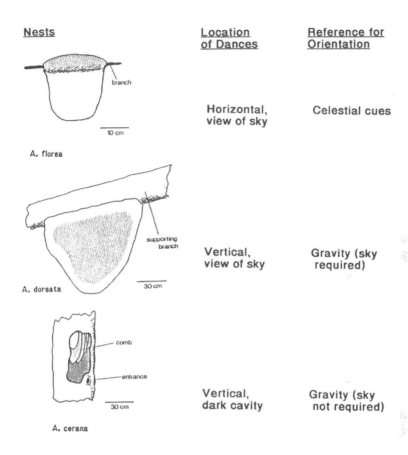

Nests	Location of Dances	Reference for Orientation
A. florea	Horizontal, view of sky	Celestial cues
A. dorsata	Vertical, view of sky	Gravity (sky required)
A. cerana	Vertical, dark cavity	Gravity (sky not required)

Figure 9.2 Comparison of nest architecture in *Apis* species. (a) *Apis cerana* (shown here) and *A. mellifera* both construct nests consisting of several parallel combs suspended from the ceiling of an enclosed cavity. (b) *A. dorsata* suspends a single large comb from tree branch or an overhanging rock; the colony protects the comb with a curtain of workers. (c) *A. florea*, which also nests in the open, suspends a single small comb from a small twig; wax wraps around the twig to form a rounded platform atop the nest. The stippled area on each nest shows the typical location of dances.

they measure their orientation relative to gravity (von Frisch, 1967). These environmental features can be used entirely independently in *A. mellifera* and other species (Dyer, 1984) and gravity is not used at all by one *Apis* species (see below).

Other features of the waggle dance have also been proposed to represent distinct traits on the strength of taxonomic variation in their expression. The sounds produced by dancers, originally thought to be an intrinsic feature of the waggling run (von Frisch, 1967; Esch et al., 1965), were found by Towne (1985) to be absent in two of the Asian bees; on the other hand, these two species exhibited exaggerated dance postures not seen in the other species. Species and subspecies variations have been observed in the transition distance, the flight distance at which directional waggle dances rather than non-directional round dances are performed (von Frisch, 1967; Towne and Gould, 1988), although Kirchner et al. (1988) raise doubts whether even round dances lack directionality. Finally, species differences have been observed in the angular precision of waggle dances (Towne and Gould, 1988).

Case Study 1:
Phylogenetic Insights Derived from
Diversity in the Direction Code

Lindauer (1956) observed that the direction codes of the Asian *Apis* species differed in various ways from that of *A. mellifera*, and that the interspecific diversity in the dances was strikingly correlated with the diversity in nesting behavior (Figure 9.2). The Asian hive bee, *A. cerana*, closely resembles *A. mellifera* both in its nest architecture (multiple combs built inside an enclosed cavity) and in its dancing behavior (using either celestial cues or gravity as a directional reference). The two other Asian species, however, build single-combed nests in the open, and both appeared to Lindauer to be wholly or partially dependent on a view of the sky for proper dance orientation. In the giant honey bee, *A. dorsata*, bees dance on the vertically hanging curtain of workers that protects the comb. Although the correlation between the solar flight angle and the orientation of the dance relative to gravity is the same as in *A. mellifera* and *A. cerana*, dancers seemed confused when Lindauer blocked the sky from view, suggesting they could not use gravity independently. In the dwarf honey bee, *A. florea*, bees dance on the rounded upper surface of the nest. Lindauer's experiments suggested that dancers are utterly dependent on a view of celestial cues, restrict their dancing to near- horizontal planes, and never orient to gravity. They became disoriented when Lindauer blocked their view of the sky. When forced to dance on the vertical sides of the nest, they seemed to do so reluctantly. Their orientation when they did dance on a vertical plane was not consistent with *A. mellifera*'s gravity-based direction code.

Based on these comparisons, Lindauer (see also von Frisch, 1967 for a review) proposed that the four *Apis* species literally reflect the major

Typical dance plane	0°	90°	90°
Gravity used	N	Y/N	Y
Sky cues used	Y	Y	Y
Nest exposure	open	open	cavity

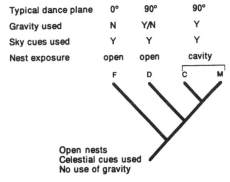

Open nests
Celestial cues used
No use of gravity

Figure 9.3 Interpretation of Lindauer's hypothesis about the evolutionary history of the waggle dance and nesting behavior of honey bees. The hypothesis suggests that *A. florea* (**F**), among extant species, most closely resembles the ancestral *Apis*, *A. dorsata* (**D**) is intermediate, and *A. cerana* (**C**) and *A.mellifera* (**M**) are most advanced. The behaviors that Lindauer observed are represented as discrete characters with alternative states expressed in different species: nests are open or enclosed in a cavity; sky cues are used for dance orientation or not; gravity is used or not (in the case of *A. dorsata*, gravity orientation was thought to be incompletely developed); the dance plane is horizontal (0°) or vertical (90°). These character sets are arrayed on the cladogram that corresponds to the phylogeny Lindauer proposed on the basis of the same characters. The cladogram is topologically identical to the cladogram proposed on the basis of independent morphological characters by Alexander (Chapter 1, this volume).

historical stages in the evolution of the dance. According to this hypothesis, *A. florea*'s behavior resembles an early stage in which foragers interacted with their nestmates atop an exposed nest, and oriented dances (perhaps crudely in the earliest stages) to celestial cues, which were also used to orient the foraging flights. *Apis dorsata* represents an intermediate stage, in which nest architecture is primitive and dancers remain partially dependent on celestial cues for orientation, but also exhibit an incipient ability to orient to gravity. *Apis cerana* and *A. mellifera* represent the most advanced stage; gravity is fully incorporated, so that their dances can be performed without a view of the sky. This presumably permitted the occupation of cavities by the ancestors of these species and the subsequent expansion of their ranges out of the tropics. Two further implications of this hypothesis are that the same mechanisms of orientation to celestial cues descended to all extant species, including those which also use gravity as a reference, and that gravity was incorporated as a reference because it allows communication in the absence of celestial cues.

The plausibility of Lindauer's historical hypothesis rests on two issues. The first is whether the actual phylogeny of the *Apis* matches the phylogeny assumed by the hypothesis. Figure 9.3 shows a cladogram corresponding to this phylogeny. Obviously, alternative phylogenies, for example one having a cavity-nesting species such as *Apis mellifera* or *A. cerana* at the basal position, would weaken Lindauer's conclusion that enclosed nests and gravity-based dances were recently-derived traits (see

Koeniger, 1976). Until recently, the phylogenetic relationships among the *Apis* species could not be unequivocally established with characters independent of those involved in dancing and nesting behavior (Ruttner, 1988). (Independent characters are necessary to avoid circularity in using the phylogeny to interpret the evolution of the behavioral characters.) However, a new analysis of morphological characters by Alexander (1991 and Chapter 1, this volume) now suggests that the phylogeny implied by Lindauer's hypothesis is basically correct. Alexander's hypothesized cladogram includes *A. andreniformis* in a clade with *A. florea*, and *A. koschevnikovi* in a clade with *A. cerana*, but topologically it is identical to the one in Figure 9.3. This result is consistent with Lindauer's assumption that the nests of the primitive *Apis* in which the dance first evolved were exposed and single-combed like those of *A. florea* and *A. dorsata*, and that primitive dancers were exposed to a view of the sky.

The second issue critical to the plausibility of Lindauer's hypothesis is whether his behavioral observations accurately reflected the pattern of character differences among honey bee species. Certain of his conclusions have indeed held up to further scrutiny: most important, it seems clear that *A. florea* dancers do not use gravity as a reference in the dance. Other conclusions, however, require revision in light of more recent discoveries.

Three findings are most important. First, *A. dorsata*, like *A. mellifera* and *A. cerana*, appears to have no difficulty orienting dances to gravity in the complete absence of celestial cues (Koeniger and Koeniger, 1980; Dyer, 1985a). Since there are no other obvious differences in the direction codes of these three species, *A. dorsata*'s dance language provides no clues about intermediate stages between dances which can be oriented to gravity and those, like *A. florea*'s, which cannot.

Second, *A. florea* dancers, in spite of their inability to orient to gravity, can orient properly without a view of celestial cues, for example on overcast days (Koeniger et al., 1982). They do so by orienting to landmarks surrounding the nest -- branches and leaves in the typical nesting site (Dyer, 1985b, 1987). To relate the positions of the nest-site landmarks to solar flight angles, both dancers and followers must sometimes see celestial cues from the nest. Given such prior experience, however, communication is apparently unimpaired in the absence of celestial cues.

Third, *A. florea* has diverged from the other species in the way it orients to celestial cues (Dyer, 1985b). The difference appears during dances on slopes (Figure 9.4a), which in *A. florea* are far commoner (Dyer, ms) than usually suggested in textbook accounts of its behavior (Wilson, 1971;

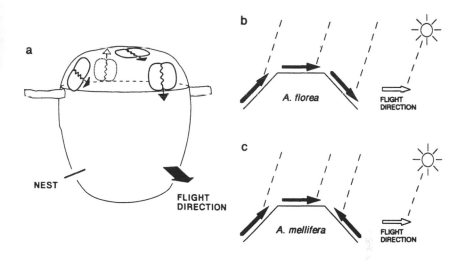

Figure 9.4 Contrasting orientation of *A. florea* and *A. mellifera* to celestial cues during dances on slopes. (a) Pattern of dance orientations typically seen on an *A. florea* nest. Dances are usually performed on moderate slopes (mostly on the side of the nest nearest the food). (b) Geometry of *A. florea*'s dance orientation on three planes, viewed from side; waggling run directions are shown with heavy arrows. (c) Pattern of orientation by *A. mellifera* dancers with view of sun on same planes, with sun in same position, and indicating the same flight direction. In *A. mellifera*, dancers align their bodies relative to the apparent azimuth in the plane of the dance. Clearly *A. florea* dancers do not do this. However, their orientation, when projected to the horizontal plane, is consistently aligned with the flight direction, suggesting that they compensate for slope and always orient to the actual solar azimuth.

Michener, 1974; Barth, 1985). In *A. mellifera, A. cerana,* and *A. dorsata,* dancers orienting to the sun on a slope align their waggling runs relative to the apparent azimuth of the sun as projected to the plane of the dance, whether the plane is horizontal or inclined (von Frisch, 1967; Edrich, 1977) (Figure 9.4c). In *A. florea,* by contrast, dancers always orient to the solar azimuth in the horizontal plane, even when dancing on steep slopes (Figure 9.4b). In other words, dancers compensate for slope in finding the solar azimuth. They compensate by rotating their heads relative to their bodies in a way that minimizes the rotation of retinal coordinates relative to the horizontal plane (Figure 9.5). Bees following dances rotate their heads in the same manner (Dyer, ms).

These three findings reinforce *A. florea*'s distinctiveness, and raise new questions about the phylogeny and adaptive significance of the character differences between it and the other *Apis* species as a group. To aid in the

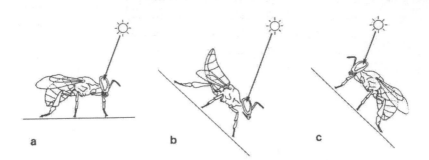

Figure 9.5 Explanation of differing orientation of *A. mellifera* and *A. florea* on slopes. In each panel dancer is shown in the middle of a waggling run indicating food in the direction of the solar azimuth. On a horizontal plane (**a**), the orientation of the two species is identical. On a slope, however, *A. mellifera* (**c**) orients to the projection of the sun in the plane of the dance, while the *A. florea* dancer (**b**), compensates for slope by rotating her head and orienting to the solar azimuth in the horizontal plane. *A. mellifera*'s orientation on the same plane would change if the sun were lower in elevation, but *A. florea*'s orientation would not change.

interpretation of the interspecific differences as they now appear to be, I have added the new results to the same cladogram shown earlier (Figure 9.6).

Note that in contrast to the original hypothesis of Lindauer (1956) and von Frisch (1967), the living *Apis* species do not exhibit clear evidence of a straightforward historical increase in the complexity of the direction code. In particular, even if we assume that the ancestral honey bee resembled *A. florea* in orienting to visual cues, but not to gravity, on an exposed nest, we cannot infer that the ability to use gravity as a reference was simply added to a primitive system of celestial orientation. Instead, the incorporation of gravity in the common ancestor of *A. dorsata*, *A. cerana* and *A. mellifera* was accompanied by a change in the way dancers respond to celestial cues when dancing on slopes.

A more fundamental point is that *A. florea*'s uniqueness renders ambiguous the evolutionary polarity of the characters that separate it from the other species, including its inability to orient to gravity. According to Lindauer's hypothesis, *A. florea*'s behavior is closest to the primitive dance language. However, with only the *Apis* species as a basis for comparison, it is equally parsimonious to conclude that the unique characteristics of *A. florea*'s dance are uniquely derived and that the characteristics of the other species represent the primitive state.

In principle, one way to resolve this sort of ambiguity is to examine homologous traits in outgroup taxa (Maddison et al., 1984), ideally in the species most closely related to the honey bees as a group. For example, one might decide which of two alternative states of a given character was primitive for the group under study by observing which state occurs in outgroup taxa. Two difficulties arise in applying this approach to the evolution of the dance language. First, the phylogenetic relationships of the honey bees and their closest relatives (bumble bees, stingless bees, and orchid bees) are still ambiguous (see Chapters 3, 4 and 5, this volume), and therefore so is the choice of the appropriate outgroups. Second, and more troublesome, only honey bees have dances, and so outgroup comparisons have to be based on behaviors performed in different contexts, raising the problem of establishing their homology.

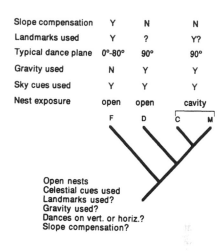

Figure 9.6 Summary of recent comparisons of dance orientation in four *Apis* species, arrayed on same cladogram shown in Figure 9.4. Question marks indicate gaps in our understanding of the behavior of extant species (top part of the figure), and ambiguities about the character states that were present in the ancestral dance.

This second difficulty may not be insurmountable. For example, in a comparative study of geotaxis, Jander and Jander (1970) observed that *A. florea's* behavior on a slope departed from that of the *Apis* species which use gravity in the dance, but resembled that of seven other species of Apoidea. *Apis florea* and the non-*Apis* bees exhibit progeotaxis, which means that the scatter around the mean uphill orientation decreases as the slope is made steeper. The other *Apis* species exhibit metageotaxis: the scatter around the mean uphill orientation stays roughly constant as slope changes. This outgroup comparison suggest that progeotaxis is primitive and metageotaxis is derived. Assuming that the mechanisms underlying metageotaxis are related to those underlying the use of gravity for dance communication (an assumption which remains to be proven), then this comparison also suggests that the use of gravity is derived rather than primitive. Further complicating this interpretation is that the proprioreceptors involved in progeotaxis in *A. florea* are not identical to those involved in progeotaxis in bumble bees (Horn, 1973, 1975), perhaps indicating independent origins of the behavior (Koeniger, 1976). In spite of the remaining ambiguities, this line of study provides a valuable example of how one might perform outgroup comparisons using orientation behaviors that may be homologous to discrete elements of the direction code.

The polarity of certain evolutionary transitions might also be resolved by establishing that one character state is ubiquitous in the ingroup (and hence primitive for that group), even if an alternative state (which could be interpreted as derived if not also ubiquitous) is normally expressed in one or more species. This situation could arise if the primitive behavioral trait were absent in some species not because the underlying neural mechanisms were absent, but because the conditions for its expression do not normally exist. (Kavanau [1990] has argued that the retention of obsolete but still accessible neural circuitry may be a widespread phenomenon.) Thus, if *A. florea* had turned out in some circumstances to orient its dances to gravity according to *A. mellifera*'s code (apparently it never does), then we might conclude, even without an outgroup comparison, that so did the (open-nesting) common ancestor of all the living *Apis* species. Similarly, if *A. mellifera, A. cerana,* and *A. dorsata* in some circumstances compensated for slope with head balancing, then we might conclude that this ability was part of the primitive dance. Finally, regarding the use of landmarks for dance orientation, I already have preliminary evidence (Dyer, in preparation) that *A. mellifera* dancers share this ability with *A. florea,* thus that it may be primitive in origin.

Now I would like to shift perspectives, and consider briefly the adaptive significance of the diversity in the traits I have discussed. The key point I would like to make is that hypotheses about adaptive significance of particular transitions from one character state to another remain unclear in part because of our uncertainty about the polarity of these transitions (Coddington, 1988). To take one example, the selective pressures we might invoke for a secondary loss of slope compensation in the common ancestor of *A. dorsata, A. cerana* and *A. mellifera* would probably differ from those invoked for a gain of this behavior in *A. florea* from an ancestor whose dance lacked it. Furthermore, the adaptive significance of a gain or loss of a trait such as slope compensation probably cannot be evaluated in isolation; since we are dealing, after all, with a component of a highly integrated behavior, the relative advantages of alternative states of one component should depend on the states of other components (e.g., whether or not dances typically occur on a vertical). Thus, a full understanding of the adaptive significance of species differences in the direction code depends upon a full understanding of the phylogeny of these differences.

Before leaving this line of study, I would like to emphasize that a modified version of Lindauer's (1956) phylogenetic hypothesis, with *A. florea* closest to the primitive *Apis* in its dance language and nesting behavior, *A. dorsata* intermediate in its nesting behavior but not in its dance language, and *A. cerana* and *A. mellifera* showing a further elaboration in nesting behavior, remains more attractive than other hypotheses that would be consistent with the phylogenetic distribution of the relevant behavioral characters. What I have hoped to do is to identify the assumptions and the types of evidence that are needed to support this or any other reconstruction of the dance language.

Case Study 2:
Adaptive Design of the Distance Code

The second major result of Lindauer's (1956) original study was his discovery of interspecific variations in the distance code of the dance. These resembled intraspecific geographic variations also discovered around the same time in *Apis mellifera* (Boch, 1957; von Frisch, 1967). Currently the most widely accepted functional explanation for these so-called dialects is that natural selection "tunes" the distance code of colonies in a given population to maximize the efficacy of communication over the range of distances that foragers typically fly (von Frisch, 1967; Gould, 1982; Punchihewa et al., 1985). When foraging range differs among populations as a result of independent factors such as habitat or body size constraints, selection is presumed to cause distance dialects to diverge. Conversely, similarities among dialects imply similar foraging ranges and convergent selection pressures. Lindauer's interspecific comparisons provided a major source of evidence for this hypothesis, but they suffer from various uncertainties. In this section I review a recent study of the Asian *Apis* species (Dyer and Seeley, 1991a) which attempted to test more directly the hypothesis that dialect is adaptively tuned to flight range.

The feature of the dance language that is at issue is the slope of the monotonically increasing function that relates the distance signal to flight distance. Of the several parameters of the dance which correlate highly with flight distance (von Frisch and Jander, 1957), the easiest to measure in the field is the average duration of a dance circuit -- the waggling run plus the return to begin the next run. The dialect function is determined for a colony by observing the dances of bees trained to an artificial feeder

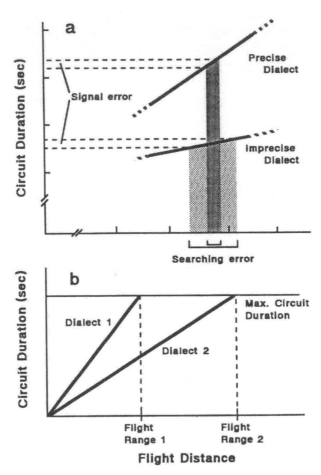

Figure 9.7 Summary of adaptive tuning hypothesis. (a) Explanation of how the precision of the dialect curve may be related to its slope. (b) If there is a fundamental constraint on maximum duration of the distance signal, and if natural selection tends to produce the steepest possible dialect curve subject to this constraint, then the dialects of populations with longer flight ranges will be less steep (Dialect 2), and presumably less precise, than those of populations with shorter flight range (Dialect 1).

set at known distances.

The adaptive tuning hypothesis rests on two assumptions (Figure 9.7). First, steeper dialect curves are assumed to be more precise, because a certain error in producing or reading the signal should translate into a relatively smaller distance error by recruit bees searching for the food (von Frisch, 1967). Second, the steepness of the curve is assumed to be limited by a constraint on maximum circuit (or waggling run) duration. For example, if flight range is long, a very steep dialect curve might result in circuit durations corresponding to the longest fight distances that are so great as to impair the ability of recruit bees to extract the distance signal: they might have difficulty in averaging successive circuits, or even in measuring a single circuit (Gould, 1982). Given some such limit on circuit duration, the optimal dialect curve is that which is the steepest (most precise) possible subject to this constraint. With the further assumption that the same constraint is shared by honey bees in different populations, then differences between populations in flight range should lead to differences in slope (Figure 9.7b).

Most previous evidence bearing on the adaptive tuning hypothesis is circumstantial, in that the foraging ranges of the populations being compared have not been directly measured (reviewed by Dyer and Seeley, 1991a). Lindauer provided two sources of evidence. First, across the *Apis* species in Sri Lanka (Figure 9.8a), he found that the steepness of the dialect curve correlated inversely with worker body size, and hence, he assumed, with foraging range. However, a relationship between body size and flight range was never verified independently. Second, Lindauer found that *A. cerana*, the species intermediate in body size and in the slope of its dialect curve, would fly a greater maximum distance to an artificial source of food than would *A. florea*, the species with the smallest body size and the steepest dialect curve. (The maximum training distance of the largest species, *A. dorsata*, was not determined.) This result supports the hypothesis only if maximum training distance correlates with maximum natural flight range, but in fact it may underestimate it to an unpredictable degree (von Frisch, 1967).

In Thailand, Dyer and Seeley (1991a) compared the dialects and flight ranges of the same species that Lindauer studied. Flight range was measured by means of forage maps (Visscher and Seeley, 1982); a colony's previously determined distance code was used to infer the distances indicated by a sample of dancers indicating natural foraging sites. Colonies of all three species were studied in a relatively undisturbed rainforest habitat during the same flowering season. This method should therefore give a fairly reliable measure of the typical flight range of each species, or at least of the typical interspecific differences in flight range. Also, the raw data allow comparison of the maximum circuit durations performed in the three species, and consequently allow a test of the assumption that the species should be similar in this respect.

The dialects of the three species in Thailand hardly differed at all (Figure 9.8b). If dialect is correlated with flight range, then their flight ranges to natural floral sources should also be similar, but we found that

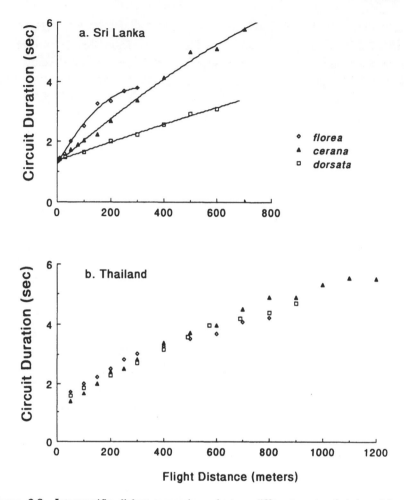

Figure 9.8 Interspecific dialect comparisons in two different parts of Asia. (a) Lindauer's (1956) data from Sri Lanka; curves are fitted by eye. (b) Dyer and Seeley's (1991) data from Thailand. Regressions on these data (not shown) were used to infer the distances flown by dancers indicating natural feeding sites in a common rainforest habitat. (Details in Dyer and Seeley, 1991a.)

they were quite different by a variety of measures (Table 9.1). *Apis dorsata*'s flight range was the longest, as Lindauer assumed it to be. On the other hand, *Apis florea*'s flight range was longer than *A. cerana*'s. This result violates the assumption that flight range scales with body size, but it accords with a wealth of other evidence that scaling relationships are reversed in these two species (Dyer and Seeley, 1987, 1991b, Chapter 11, this volume). Whatever the cause of this pattern, these comparisons failed to reveal any association between dialect and flight range. The same conclusion is supported by a parallel study recently carried out in Sri

Table 9.1 Flight ranges inferred from dances to natural food sources in four *Apis* species, comparing medians and 95th percentiles of overall distributions (N = number of observations), and, as a measure of maximum flight distances, means of upper 2% of measured circuit durations and inferred flight distances.

| Species | N | Distribution of Inferred Flight Distances | | Mean of Upper 2% of Distribution | |
		50% (m)	95% (m)	Flight Distances (km)	Circuit Duration (s)
Apis florea[a]	380	268	1323	11.2 ± 2.5 (n=8)	30.3 ± 6.3
Apis cerana[a]	912	195	905	1.9 ± 0.4 (n=18)	7.6 ± 1.1
Apis dorsata[a]	501	863	3810	12.0 ± 6.1 (n=11)	28.9 ± 13.3
Apis mellifera[b]	1871	1650	6000	7.9 ± 0.9 (n=37)	~9.5

[a]Data from Dyer and Seeley (1991)
[b]Data from Visscher and Seeley (1982)

Lanka (Punchihewa et al., 1985), although these authors interpreted their results otherwise (see Dyer and Seeley, 1991a, for discussion).

The data from Thailand (Table 9.1) also undermine a key assumption of the adaptive tuning hypothesis, namely that there is a common constraint on maximum circuit duration. The maximum circuit durations in *A. florea* and *A. dorsata* were similar, but both were 3-4 times as great as those observed in *A. cerana* and *A. mellifera*. Hence, the adaptive tuning hypothesis is further weakened as a general explanation for interspecific differences and similarities in the dialects. At best the relationship between dialect and foraging ecology must be more subtle than previously supposed.

Our finding that maximum circuit duration was similar for the two open-nesting species, and similar for the two cavity-nesting species, suggests one possible way to rescue the adaptive tuning hypothesis. This pattern is striking because it joins a growing body of diverse evidence that behavioral and physiological traits tend to converge in *Apis* species which share the same basic nest architecture (Dyer and Seeley, 1987, 1991b, Chapter 11, this volume). Perhaps there are differing constraints on maximum circuit duration according to whether dances are usually performed on an exposed nest, where vision may be used by dance followers, or in a cavity, where only tactile and acoustic information can be used (Towne, 1985; see below). If so, then the adaptive tuning hypothesis may be appropriate to explain similarities and differences among species that share similar nesting behavior. In qualitative support of this possibility, *A. florea* and *A. dorsata* have similar dialects and similar (maximum) flight ranges in Thailand; comparing the cavity-nesting species, *A. cerana* in Thailand has a steeper dialect curve and a shorter flight range than *A. mellifera* studied in a northern temperate forest (Table 9.1).

If dialects are the result of optimal tuning of slope subject to a constraint on circuit duration that varies with nesting behavior, then these results provide an interesting new historical perspective on the dialect curve as an adaptively designed property of the dance language. The evolution of dialects is usually considered without reference to history, as if the ancestral shape of the dialect curve were irrelevant. But if exposed nests represent the ancestral state (Figures 9.3, 9.6), then so might dances with relatively long maximum circuit durations and relatively steep dialect curves for a given foraging range. By extension, the move into cavities might have been accompanied by a reduction in the maximum circuit duration, and a reduction in the slope and hence the potential precision of the dialect curve.

As intriguing as this line of reasoning is, it is no better supported than the original adaptive tuning hypothesis. Additional joint comparisons of flight range and dialect are urgently needed, as are direct tests of the assumptions that maximum circuit duration constrains the shape of the dialect curve in some way, and that steeper dialect curves are more precise.

Case Study 3:
History and Function of the Dance Sounds

Acoustic signals were first discovered in honey bee dances by Esch (1961) and Wenner (1962). There is increasing evidence that they are crucial for communication in *A. mellifera* (Michelsen et al., 1987). Most early speculations about the evolution of the dance (e.g., von Frisch, 1967) took for granted that sounds are an integral and ancient part of the waggling run. In fact, some stingless bees foragers produce airborne sounds when performing agitated non-oriented movements in the nest between flights to the food, a discovery which led to the suggestion that communicative "dances" containing sounds antedated the separation of stingless bees and honey bees (Esch et al., 1965). This hypothesis depends in part on the assumption that the most recent common ancestor of these groups was social, but this question is still being debated (Chapters 3, 4, and 5, this volume). Moreover, Towne (1985) has proposed that, among the *Apis*, only the cavity-nesting species produce dance sounds, raising the possibility that they have been independently derived within honey bees.

Towne recorded very similar dance sounds in the two cavity-nesting species, *A. cerana* and *A. mellifera*, but was unable to record sounds in the two open-nesting species *A. florea* and *A. dorsata*. He suggested that sounds evolved to intensify the signal in dances performed in darkness. Intriguingly, dancers in the two open-nesting species adopt a posture during waggling runs that is visually conspicuous, at least to human observers: the dancer's wings are spread wide (rather than held close to the body as in the cavity-nesters) and the abdomen waggles dorso-ventrally as well as from side to side. Lindauer (1956) suggested that these apparently disordered movements represented a less perfected ancestral state of the dance. Towne proposed that they are designed for visual communication of the dance signal in these species.

More recently, improved recording techniques have been used to detect dance sounds in both of the open-nesting species (W. Kirchner and F. Dyer, unpublished observations). This weakens Towne's hypothesis that sounds are less important for dance communication in the open-nesting species than in the cavity-nesting species. On the other hand, Towne's suggestion that visual signals are more important in the open-nesting species is still highly plausible. If this is verified experimentally, then it offers new insights (and raises new questions) about the phylogeny of the dance language. Given the likelihood that the primitive *Apis* nested in the open, visual signals presumably played a role in the primitive honey bee dance, but were lost with the move into cavities. One question this raises is whether other changes necessarily accompanied this transition.

Conclusions

Darwin considered an important test of his theory of natural selection to be its ability to explain the evolution of such "organs of extreme perfection and complication" as the vertebrate eye or, to take one of his behavioral examples, comb construction by the honey bee. Darwin's strategy for formulating such explanations has clearly inspired much of the theorizing about the evolution of the honey bee's dance language, a trait as striking in its complexity as the ones he considered. Much has been learned, for example, by adopting the working hypothesis that the various components of the dance stand in the same relationship to the whole as do the eye's "...inimitable contrivances for adjusting the focus to different distances, for admitting different amounts of light, and for the correction of spherical and chromatic aberration..." (Darwin, 1859). In other words, components of the dance language such as the distance code, or the particular aspects of the direction code, have histories and functions which to a certain extent can be analyzed separately from one another, thus allowing us to propose "how possibly" (Brandon, 1990) the whole might have arisen.

Furthermore, studies of the dance have underscored the value of interspecific comparisons as a source of knowledge about the origin and modification of behavioral traits. Ambiguities inherent in describing behavioral characters, identifying homologies, and ascribing functional significance, pose major hurdles to understanding behavioral evolution. The case studies I have discussed suggest that the way past these hurdles may sometimes be found by analyzing the behaviors of different species in ever more detail, and by interpreting behavioral diversity in a historical context based on independent information about phylogeny.

Acknowledgments

The author's research has been supported by the Smithsonian Institution (Special Foreign Currency grants), a Fulbright Fellowship, the National Science Foundation (BNS-8405962 and BSR-8918932) and Michigan State University (AURIG).

Literature Cited

Alexander, B. A. 1991. Phylogenetic analysis of the genus *Apis*. *Ann. Entomol. Soc. Am.* 84:137-149.

Barth, F. G. 1985. *Insects and flowers*. Princeton University Press: Princeton, NJ.

Boch, R. 1957. Rassenmäßige Unterschiede bei den Tänzen der Honigbiene (*Apis mellifica* L.), *Z. vgl. Physiol.* 40:289-320.

Brandon, R. N. 1990. *Adaptation and environment*. Princeton University Press: Princeton, NJ.

Coddington, J. A. 1988. Cladistic tests of adaptational hypotheses. *Cladistics* 4:3-22.

Dawkins, M. 1983. The organisation of motor patterns. In *Animal behavior (Vol. 1): causes and effects.*, T. R. Halliday and P. J. B. Slater, eds. Blackwell: Oxford. Pp. 75-99.

Dyer, F. C. 1984. *Comparative studies of the dance language and orientation of four species of honey bees.* Thesis, Princeton University.

------. 1985a. Nocturnal orientation by the Asian honey bee, *Apis dorsata. Anim. Behav.* 33:769-774.

------. 1985b. Mechanisms of dance orientation in the Asian honey bee *Apis florea. J. Comp. Physiol. A* 157:183-198.

------. 1987. New perspectives on the dance orientation of the Asian honey bees. In *Neurobiology and behavior of honeybees.*, R. Menzel and A. Mercer, eds. Springer: Berlin. Pp. 54-65.

Dyer, F. C. and T. D. Seeley. 1987. Interspecific comparisons of endothermy in honey-bees (*Apis*): deviations from the expected size-related patterns. *J. Exp. Biol.* 127:1-26.

------. 1989a. On the evolution of the dance language. *Am. Nat.* 133:580-590.

------. 1989b. Orientation and foraging in honeybees. In *Insect flight.*, G. J. Goldsworthy and C. H. Wheeler, eds. CRC Press: Fort Lauderdale. Pp. 205-230.

------. 1991a. Dance dialects and foraging ranges in three Asian honey bee species. *Behav. Ecol. Sociobiol.* in press.

------. 1991b. Nesting behavior and the evolution of worker tempo in four honey bee species. *Ecology* 72:156-170.

Edrich, W. 1977. Interaction of light and gravity in the orientation of the waggle dance of honey bees. *Anim. Behav.* 25:342-363.

Esch, H. 1961. Über die Schallerzeugung beim Werbetanz der Honigbiene. *Z. vgl. Physiol.* 45:1-11.

Esch, H., I. Esch and W. E. Kerr. 1965. Sound: an element common to communication of stingless bees and to dances of the honey bee. *Science* 149:320-321.

Frisch, K. v. 1967. *The dance language and orientation of bees.* Harvard University Press: Cambridge, MA.

Frisch, K. v. and R. Jander. 1957. Über den Schwänzeltanz der Bienen. *Z. vgl. Physiol.* 40:239-263.

Gould, J. L. 1982. Why do honey bees have dialects? *Behav. Ecol. Sociobiol.* 10:53-56.

Horn, E. 1973. Die Verarbeitung des Schwerereizes bei der Geotaxis der höheren Bienen (Apidae). *J. Comp. Physiol.* 82:397-406.

------. 1975. Mechanisms of gravity processing by leg and abdominal gravity receptors in bees. *J. Insect Physiol.* 21:673-679.

Jander, R. and U. Jander. 1970. Über die Phylogenie der Geotaxis innerhalb der Bienen (Apoidea). *Z. vergl. Physiol.* 66:355-368.

Kavanau, J. L. 1990. Conservative behavioral evolution, the neural substrate. *Anim. Behav.* 39:758-767.

Kirchner, W. H., M. Lindauer and A. Michelsen. 1988. Honey bee dance communication: Acoustical indication of direction in round dances. *Naturwissenschaften* 75:629-630.

Koeniger, N. 1976. Neue Aspekte der Phylogenie innerhalb der Gattung *Apis. Apidologie* 7:357-366.

Koeniger, N. and G. Koeniger. 1980. Observations and experiments on migration and dance communication of *Apis dorsata* in Sri Lanka. *J. Apic. Res.* 19:21-34.

Koeniger, N., G. Koeniger, R. K. W. Punchihewa, Mo. Fabritius and Mi. Fabritius. 1982. Observations and experiments on dance communication of *Apis florea* in Sri Lanka. *J. Apic. Res.* 21:45-52.

Krebs, J. R. and R. H. McCleery. 1984. Optimization. In *Behavioural ecology: an evolutionary approach.*, J. R. Krebs and N. B. Davies, eds. Blackwell: Oxford. Pp. 91-121.

Lauder, G. V. 1986. Homology, analogy, and the evolution of behavior. In *Evolution of animal behavior: paleontological and field approaches.*, M. H. Nitecki and J. A. Kitchell, eds. Oxford University Press: Oxford. Pp. 9-40.

Lindauer, M. 1956. Über die Verständigung bei indischen Bienen. *Z. vgl. Physiol.* 38:521-557.

------. 1985. The dance language of honeybees: the history of a discovery. In *Experimental behavioral ecology and sociobiology.*, B. Hölldobler and M. Lindauer, eds. G. Fischer Verlag: Stuttgart (*Fortschritte der Zoologie* 31). Pp. 129-140.

Maddison, W. P., M. J. Donoghue and D. R. Maddison. 1984. Outgroup analysis and parsimony. *Syst. Zool.* 33:83-103.

Michelsen, A., W. F. Towne, W. H. Kirchner and P. Kryger. 1987. The acoustic near field of a dancing honeybee. *J. Comp. Physiol.* 161:633-643.

Michener, C. D. 1974. *The social behavior of the bees.* Harvard University Press: Cambridge, MA.

Punchihewa, R. W. K., N. Koeniger, P. G. Kevan and R. M. Gadawski. 1985. Observations on the dance communication and natural foraging ranges of *Apis cerana, Apis dorsata,* and *Apis florea* in Sri Lanka. *J. Apic. Res.* 24:168-175.

Ruttner, F. 1988. *Biogeography and taxonomy of honeybees.* Springer: Berlin.

Seeley, T. D. 1985. *Honeybee ecology.* Princeton University Press: Princeton, NJ.

------. 1989. The honey bee colony as a superorganism. *Am. Sci.* 77:546-553.

Towne, W. F. 1985. Acoustic and visual cues in the dances of four honey bee species. *Behav. Ecol. Sociobiol.* 16:185-187.

Towne, W. F. and J. L. Gould. 1988. The spatial precision of the honey bees' dance communication. *J. Insect Behav.* 1:129-155.

Visscher, P. K. and T. D. Seeley. 1982. Foraging strategy of honeybee colonies in a temperate deciduous forest. *Ecology* 63:1790-1801.

Wehner, R. 1984. Astronavigation in insects. *Annu. Rev. Entomol.* 29:277-298.

Wenner, A. M. 1962. Sound production during the waggle dance of honey bees. *Anim. Behav.* 10:79-95.

Wilson, E. O. 1971. *The insect societies.* Harvard University Press: Cambridge, MA.

Winston, M. L. 1987. *The biology of the honey bee.* Harvard University Press: Cambridge, MA.

10

Diversity in
Apis Mating Systems

Gudrun Koeniger

In Africa and Europe only one species of *Apis* -- *Apis mellifera* --
occurs naturally, and its mating behaviour seems to be rather uniform over
all its range. On the other hand, several species of *Apis* are sympatric in
Asia: *Apis cerana*, *A. koschevnikovi*, *A. florea*, *A. andreniformis* and *A.
dorsata* (including *A. d. laboriosa*). To avoid interspecific mate competition the mating systems of symaptric species must be divergent.
Comparative studies of mating systems in the Asian honey bee species and
Apis mellifera might give us some ideas about which aspects of the mating
system to increase the fitness of the honey bee colony and which
characters ensure reproductive isolation (Koeniger and Koeniger, 1990b).

Sex Pheromones

There is apparently little difference in the pricipal components of the
sex pheromones of the *Apis* species which have been examined. The
mandibular glands in all species contain 9-oxodecenoic acid (Shearer et al.,
1970; Sannasi et al., 1971). This component proved to be the main sex
attractant in *A. mellifera* (Gary, 1962; Pain and Ruttner, 1963) and is also
active in *A. cerana*, *A. florea* and *A. dorsata*. There are no data available
on the sex pheromones of *A. andreniformis* or *A. koschevnikovi*.
Additional sex pheromones were found in the secretion of the tergite
glands in *A. mellifera* (Renner and Vierling, 1977); other species were not
tested.

There is also no difference in the attractiveness of ethanol extracts of
A. florea, A. dorsata, A. cerana or *A. mellifera* queens to *A. mellifera*
drones (Butler et al., 1967; Shearer et al., 1970; Sannasi et al., 1971).
Thus interspecific pheromonal attraction, which could disturb natural
mating, is possible. Other features of the mating system must prevent such
complications.

Flights

Drones

In all species drones leave the colony only for a limited time each day.
Drones fly from the nest in 3 situations: cleansing flights; orientation and
mating flights; and swarm flights.

Drones often join the cleansing flights of worker bees. These flights
occur mostly around noon and last for 5 - 10 minutes in *A. mellifera, A.
florea* and *A. cerana* (G. Koeniger, unpublished observations) and in *A.
dorsata* (Mardan, 1989 pers. comm.). Individual flight duration during
cleansing flights were not observed.

Orientation and mating flights occur within the actual drone flight time,
which is independent of the worker flight. The individual flight time for
orientation normally lasts for 1-2 min (Drescher, 1968; Berg, 1989).
Individual flight time for mating lasts 20-30 min in *A. m. carnica* (Berg,
1989), 10 to 15 min in *A. cerana indica* (Punchihewa, pers. comm.) and
15 to 25 min in *A. florea* (unpublished observations on two colonies, 12
marked drones, 52 flights).

In naturally sympatric species the drone flight time is species specific
(Figure 10.1). This is one aspect of an interspecific reproductive barrier
(Koeniger and Wijayagunesekara, 1976; Koeniger et al., 1988).

Intraspecific variation in flight time in different locations has also been
reported (Figure 10.1). Drones of *A. cerana indica* fly from 15.30 h -
17.00 h in Sri Lanka (Koeniger and Wijayagunesekara, 1976; Punchihewa
et al., 1990a), from 14.00 h - 15.30 h in Borneo (Koeniger et al., 1988),
from 14.30 h - 16.00 h in Bangkok Thailand (G. Koeniger, unpublished
observations). In Japan, drones of *A. cerana japonica*, the only indigenous
honey bee species, have a longer flight period. They fly from 13.15 h -
16.15 h (Yoshida and Saito, 1990). Flight time may also differ slightly
among *A. mellifera* races or subspecies at the same location. This was true
for Africanized and European honeybees (Rowell et al., 1986). However
no difference in flight time was found between *A. m. carnica* and *A. m.
ligustica* (Drescher, 1968; Koeniger et al., 1989a). Drone flight time of

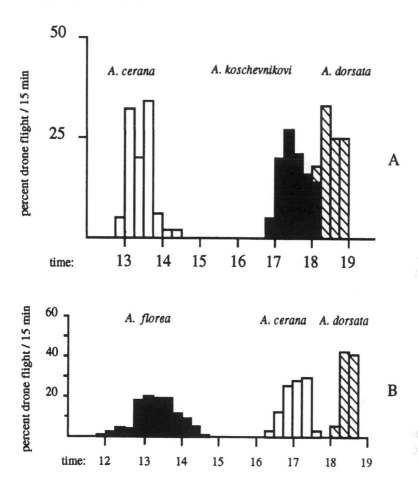

Figure 10.1 Drone flight time in (A) Sabah, Borneo and (B) Sri Lanka (after Koeniger et al., 1988; Koeniger and Wijayagunesekera, 1976).

the same colony also shows a time shift dependent on locations, climate conditions and time of the year (Taber, 1964).

In regions where only one species naturally exists, mating flight time lasts between 3 and 4 hours (Japan and Austria), while in regions with sympatric species it is only 1 to 1 1/2 hours. Introducing a non-native species may render matings more difficult. In Germany drones of introduced *Apis cerana* from China fly at the same time as drones of *A. mellifera carnica* (Ruttner et al., 1972). In Japan the native *A. cerana japonica* drones start flying before the drone flight of the imported *A. mellifera* has ended (Yoshida and Saito, 1990).

Table 10.1 Number of spermatozoa (in millions) found in the drone seminal vesicle and queen spermatheca in six *Apis* species (after Koeniger et al., 1990a).

Species	Vesicle	Spermatheca
A. mellifera	10.0	5.0
A. cerana	01.2	1.4
A. koschevnikovi	?	?
A. dorsata	02.5	3.6
A. florea	00.4	1.4
A. andreniformis	00.1	1.1

Queens

Queens normally fly during the peak of drone flight, but in *Apis florea* virgin queens were also observed to join the cleansing flight of workers, before noon. Before mating, queens often perform orientation flights, which last about 2 minutes. *Apis florea* queens were mated between 14.04 h and 15.32 h. The flight duration for successful mating flights was 21 to 30 minutes (Koeniger et al., 1989b).

Queens of *A. cerana* flew between 13.30 h and 15.30 h in Poona, India and mating occurred between 14.00 h and 15.00 h (Woyke, 1975). In Sri Lanka queens were mated between 16.15 h and 16.55 h (Punchihewa et al., 1990a). In Japan the queens flew between 13.15 h and 17.15 h and were mated between 14.45 h and 16.00 h (Yoshida and Saito, 1990). Queens as well as drones fly for a much longer period in Japan, where *A. cerana* is the only native honeybee, than in locations where other honey bee species naturally occur. Mating flight per queen lasted around 10 minutes in Sri Lanka, 20 to 40 min in Poona, India and between 16.5 and 37.4 min in Tokyo, Japan.

In Austria *A. mellifera carnica* and *ligustica* queens were mated between 13.59 h and 16.01 h and 15.24 h and 16.16 h respectively. Drones of both species were caught on the drone congregation area (DCA; see below) in equal numbers between 14.30 h and 17.00 h (Koeniger et al., 1989a). Individual successful mating flight time was 15 min in *A. m. carnica* and 12 min in *A. m. ligustica* in Austria. In Venezuela Africanized queens and drones are better synchronized in their mating flights than the European race (Hellmich and Collins, 1990). The average flight time for successful matings was 18.4 min for Africanized queens and 26.3 min for European queens.

Table 10.2 Data on the spermatheca of six species of *Apis* (after Koeniger et al., 1990a).

Species	Diameter (mm)	Volume (mm³)	No sperm (X 10⁶)	Concentration (10⁶/mm³)
A. mellifera	1.14	0.78	4.73	6.12
A. cerana	0.98	0.48	1.35	2.79
A. koschevnikovi	?	?	?	?
A. dorsata	1.1	0.70	3.67	5.24
A. florea	0.8	0.27	1.15	4.41
A. andreniformis	0.8	0.27	1.04	3.83

Place of Mating

Drones of *A. mellifera* congregate in free flight over certain places, called drone congregation areas (DCAs; see below), typically in areas without high vegetation. In Austria these places were continuously visited over more than 20 years. The height of flight differs according to climatic conditions (Ruttner and Ruttner, 1963) and also according to the race (Koeniger et al., 1989a). Drones of *A. cerana* colonies introduced to Germany were caught on the same DCA as *A. mellifera* drones (Ruttner et al., 1972). In Japan a few drones of *A. cerana japonica* were caught on DCAs of introduced *A. mellifera* during the *mellifera* drone flight. However, after the cessation of *mellifera* drone flight, no additional *cerana* drones were caught on the DCA, even though the peak of *cerana* drone flight occured after *mellifera* drone flight (Yoshida and Saito, 1990). This seems to indicate that *cerana* drones normally meet elsewhere.

Drones of native *A. cerana* in Sri Lanka assemble in the canopy of trees; thus the height of flight is dependent on the height of the trees (Punchihewa et al., 1990b). The same is true in Malaysia (Mardan, pers. comm.). *Apis cerana japonica* drones have been reported to congregate above the tops of native chestnut trees (Seito Fujiwara, pers. comm. to editor). At Kilimanjaru, Tanzania drones of native *A. mellifera* were observed within the canopy of trees (Ruttner, pers. comm.). The question this arises whether the place of mating is dependent on local conditions such as climate and predators or if it is an additional factor for interspecific reproductive isolation.

No data on DCAs are known for the other *Apis* species.

Flight Range

The flight range of *A. mellifera* drones is reported to vary between 1 and 5 km, and that of queens between 1 and 2 km (Ruttner and Ruttner, 1966; Woyke, 1960; Taylor and Rowell, 1988). In isolated areas the mating distance can range up to 16 km (Peer, 1957). Mating distance of *A. cerana* in Sri Lanka seems not to exceed 500 m (Punchihewa et al., 1990b). There are no data for the other *Apis* species.

Sperm Numbers in Drones and Queens

The number of sperm produced by individual drones ranges from 0.1 million in *A. andreniformis* to 10 million in *A. mellifera* (Table 10.1; Koeniger et al., 1990a). The interspecific differences in drone body weight are not as large, ranging from 85 mg in *A. florea* to 200 mg in *A. mellifera*.

The queens store between 1 and 6 million spermatozoa in their spermathecae (Table 10.2). The size of the spermatheca ranges from 0.27 mm^3 to 0.78 mm^3. The concentration of spermatozoa in the spermatheca shows less differences, between 2.8 million spermatozoa/mm^3 and 6.1 million/mm^3 (Koeniger et al., 1990a). The number of spermatozoa seems to be related to the size of the spermatheca.

Sperm Transfer

There are basically two different modes of sperm transfer in *Apis*: injection of spermatozoa into the oviducts, followed by movement of spermatozoa to the spermatheca, or injection of sperm directly into the spermatheca.

In *A. mellifera* and *A. cerana*, drones inject their spermatozoa into the oviducts of the queen (Figure 10.3). The queens can hold the sperm of up to 25 drones in the oviducts (Woyke, 1960, 1975). Over the next 24 hours about 5 million of the 120 million spermatozoa in an *A. mellifera* queen's oviducts will reach the spermatheca. In *A. cerana* an average of 1.3 million of the 12 million spermatozoa in the oviducts reach the spermatheca. Drones of *A. florea* transfer their sperm directly into the spermatheca (Koeniger et al., 1989b).

The anatomy of the endophalli (Simpson, 1960; Ruttner, 1988) seems to be adapted to the different mechanisms of sperm transfer. In *A.*

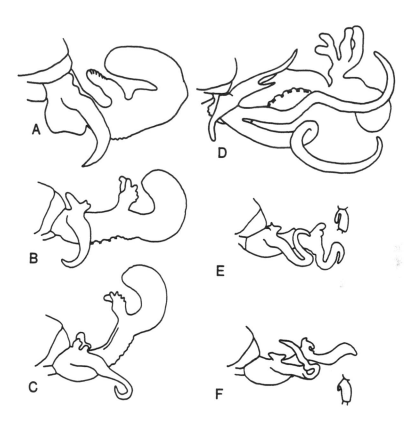

Figure 10.2 Endophalli of six *Apis* species, and metatarsi of *A. florea* and *A. andreniformis* male hind leg. A = *A. mellifera*, B = *A. cerana*, C = *A. koschevnikovi*, D = *A. dorsata*, E = *A. florea*, F = *A. andreniformis*.

mellifera sperm is transferred when the slender but short cervix is everted (Figure 10.3A), while in *A. florea* sperm transfer seems to occur only after eversion of the whole slender and long endophallus (Figure 10.3B). The endophallus of *A. andreniformis* also ends in fine tip (Wongsiri et al., 1990; Figure 10.2) and the drone has very few spermatozoa (Table 10.1). Thus one can expect direct sperm transfer in *A. andreniformis* as well. *Apis koschevnikovi* (Tingek et al., 1988) drones have the same type of endophallus as *A. mellifera* and *A. cerana* (Figure 10.2). The endophallus of *A. dorsata* has a slender and elongated cervix and small bulb (Simpson, 1970; McEvoy and Underwood, 1988; Koeniger et al., 1990b). The mechanism of sperm transfer in this species is still not clear.

Number of Matings

Drones become paralyzed during the eversion of the endophallus, and die shortly after copulation. Thus drones of all *Apis* species are monogamous.

The number of times queens mate can be estimated by counting the spermatozoa the queens receive during one mating flight and to observe the number of mating flights which a queen undertakes. This has been done for 3 species: *A. mellifera, A. cerana* and *A. florea* (Tryasko, 1956; Woyke, 1960, 1975; Koeniger et al., 1989b). According to this method the number of matings for the first two species is about 10 to 12, while for *A. florea* it is 3 to 4. Comparing the sperm numbers of queens and drones in the other species (Table 10.1) reveals that in all species queens are polyandrous (Koeniger and Koeniger, 1990a; Koeniger et al., 1990a).

For *A. mellifera* the frequency of diploid drones (Adams et al., 1977) and the allelic frequency of malate dehydrogenase (Cornuet et al., 1986) in the progeny of queens have also been used to estimate the number of matings. These methods revealed averages of 17 matings per queen in Brazilian bees and 12.4 matings in *A. mellifera mellifera*.

Significance of Mucus

Copulation

In both *A. cerana* and *A. mellifera* drones produce white mucus and an orange sticky cornual secretion (Koeniger et al., 1990c) that form the mating sign found on queens after copulation. In contrast, *A. florea* queens had no mating sign when they returned. The mucus glands of this species are very small (Koeniger et al., 1989b).

In *A. mellifera* the mucus presumably has an additional function. During mating the drone becomes paralysed after eversion of only half of the endophallus. He swings backwards before sperm transfer has occurred, the pair still hanging together and the queen still flying. In this position the membranous endophallus is very firm, because it is filled with mucus and the chitinized plates. The cornua, with their sticky secretion, increase the firmness and the hold. Thus the endophallus is inserted into the queen like "a cork in a bottle". The queen then seems to press the endophallus and sperm transfer occurs. After full eversion of the endophallus the drone falls to the ground, leaving mucus, the chitinized plates and the cornual secretions in the queen, forming the mating sign (Koeniger, 1985, 1986).

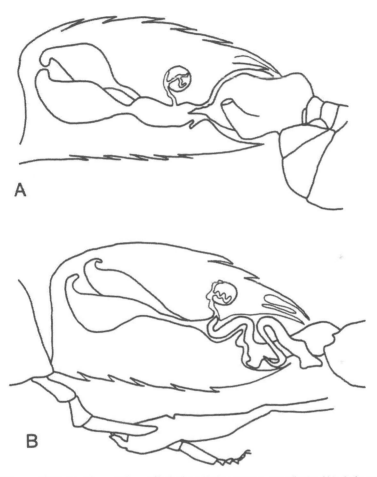

Figure 10.3 Position of the endophallus during sperm transfer in (**A**) *Apis mellifera* and (**B**) *Apis florea* (hypothesized).

Drones of *A. florea* and *A. andreniformis* have a forceps-like appendage on the metatarsus of the hind legs. With this "thumb" they anchor themselves to the hind legs of the queen (Ruttner, 1988), supported again by the sticky cornua pressed into the sting chamber. Thus, the pair stays connected until the queen turns her leg in such a way that the drone is released. Because of this mechanism, mucus is not necessary for the mating process and mucus glands are tiny in these species.

The Mating Sign

The other function of mucus is to form part of the mating sign observed in *A. mellifera* and *A. cerana*. Judging from the size of the mucus glands in *A. dorsata* and *A. koschevnikovi*, drones of these species should be able to produce a mating sign as well, but it has not yet been observed. The mating sign fills the sting chamber and bursa copulatrix of the queen and has been thought for a long time to function as mating plug which can prevent or complicate subsequent matings. However, in *A. mellifera* the mating sign is removed easily by the next drone with a special set of hairs on his endophallus (Figure 10.3). Instead of complicating multiple matings, a drone marks the queen after transfer of his own sperm with a conspicuous, coloured mating sign: the mucus is white and uv-reflecting and the cornual secretion contains a fluorescent orange pigment. Thus the mating sign facilitates the identification of the queen by the following drones. This will reduce the mating-flight time without reducing the number of matings (Koeniger, 1990).

Summary and Discussion

There are many common features in the mating system of all species:

1.   9-oxo-2-decenoic acid is the main component of the sex phero-mone.
2.   The copulatory organ is a membranous endophallus, which is everted by hemolymph pressure. This kind of endophallus is unique in Hymenoptera.
3.   Drones are monogamous, queens are polyandrous.
4.   Queens and drones leave their colony and mate in free flight. Flight distance is known only for *A. mellifera* and *A. cerana*, but the durations of flights made by individual drones suggest that all *Apis* species mate outside the immediate vicinity of their colonies.

Within this framework there are considerable differences between the species. The common sex pheromone could delay or even prevent natural mating if drones of all sympatric species were attracted to one queen. But the time of mating flights is different between sympatric species, thus giving an ethological isolation. The differences in drone congregation areas between *A. mellifera* and *A. cerana* indicate that there might be also be species-specific differences in the characteristics of DCAs, but more data are needed to prove this suggestion.

Drones of different *Apis* species have different mating strategies. *Apis mellifera* drones produce a high surplus of spermatozoa, more than the spermatheca can hold. The sperm are injected into the oviducts. They also produce a considerable amount of mucus, amounting to about 5% of their body weight (Berg, 1988). These drones are big and heavy compared to the queen. To ensure multiple mating the queen rejects more than 9/10 of the sperm she receives. This has led to a male cooperative behaviour by marking the queen with a "mating sign" to ensure multiple mating in a short flight period.

In *A. cerana* and *A. dorsata* drones have fewer spermatozoa than *A. mellifera* drones, less, in fact, than a queen stores in the spermatheca (Table 10.1). The mucus glands are about half as big. The drones are smaller than queens. *A. cerana* drones, like *A. mellifera* drones, mark the queen with a mating sign.

Drones of *A. florea* and *A. andreniformis* can inject their spermatozoa directly into the spermatheca. Because of this different mechanism in sperm transfer there are principal differences in the shape of the endophalli. To ensure multiple mating they produce much fewer spermatozoa than the queen can store. Also they produce very little mucus. To hold the queen they have a special organ at the metatarsus of the hind legs. Because of these features their abdomen is small compared to the other species and also compared to the total weight of the drone. Thus their ergonomics seem to be more efficient than in the other types of drones.

The investment of the colonies in drones follows different strategies. *Apis florea* and *A. andreniformis*, with tiny mucus glands and a small amount of semen, are not only more economic in flight but are also more economic to produce. The other extreme, drones of *A. mellifera*, with surplus sperm and huge mucus glands, requires a bigger investment.

Multiple mating by the queen and the occurence of matings outside of the vicinity of the colony is a general character in all species. This behaviour increases the survival risks for the queen, but it ensures avoidance of inbreeding and increases the genetical variability of the colony. This strategy seems to be a crucial feature which increases the fitness of a colony in all *Apis* species.

The divergence in flight time and flight pattern at the DCAs seem to be adaptations to climate, predators and to interspecific reproductive isolation.

Literature Cited

Adams, J., E. D. Rothmann, W. E. Kerr and Z. L. Paulino. 1977. Estimation of sex alleles and queen matings from diploid male frequencies in a population of *Apis mellifera*. *Genetics* 86:583-596.

Berg, S. 1988. Paarungsflug der Drohnen (*Apis mellifera* L.): Flugdauer, Flug-geschwindigkeit und Masseverlust in Abhängigkeit von unterschiedlichen Körpergrößen. Diplomarbeit, Institut für Bienenkunde, FB Biologie, Universität Frankfurt.

------. 1989. Größenabhängige Flugdauer beim Paarungsflug der Drohne. In *The flying Honeybee.*, W. Nachtigall, ed. Biona Report 6:43-50

Butler, C.G., D. H. Calam and R. K. Callow. 1967. Attraction of *A. mellifera* drones by the odours of the queens of two other species of honeybees. *Nature* 213:423-424.

Cornuet, J.-M., A. Daoudi and C. Chevalet. 1986. Genetic pollution and number of matings in a black honeybee (*Apis mellifera mellifera*) population. *Theor. Appl. Genet.* 73:223-227

Drescher, W. 1968. Die Flugaktivität von Drohnen der Rasse *A. mellifera carnica* L. und *A. mellifera ligustica* L. in Abhängigkeit von Lebensalter und Witterung. *Z. Bienenforsch.* 9:390-409.

Gary, N. E. 1962. Chemical mating attractants in the queen honeybee. *Science* 136:773-774

Hellmich, R. L. and A. M. Collins. 1990. A nuptial disadvantage for European honeybees in Venzuela. In *Social insects and the environment, Proceedings of the 11th International Congress of IUSSI.*, G. K. Veeresh, B. Mallik and C. A. Virakthamath, eds. Oxford & IBH Publishing Co. Ltd.: Bombay, Delhi. Pp. 101-102.

Koeniger, G. 1985. Funktionsmorphologische Befunde bei der Kopulation der Honigbiene. *Apidologie* 15:189-104.

------. 1986. Mating sign and multiple mating. *Bee World* 67:141-159.

------. 1990. The role of the mating sign in honeybees *Apis mellifera* L.: Does it hinder or promote multiple mating? *Anim. Behav.* 39:444-449.

Koeniger, G. and N. Koeniger. 1990a. Unterschiedliche Genese der Polyandrie bei Apis-Arten. In *Verh. Dtsch. Zool. Ges.*, Gustav Fischer Verlag: Stuttgart. Pp. 83,143.

------. 1990b. Evolution of reproductive behavior in honey bees. In *Social insects and the environment, Proceedings of the 11th International Congress of IUSSI.*, G. K. Veeresh, B. Mallik and C. A. Virakthamath, eds. Oxford & IBH Publishing Co. Ltd.: Bombay, Delhi. Pp. 101-102.

Koeniger, G., N. Koeniger, M. Mardan, R. W. K. Punchihewa and G. Otis. 1990a. Numbers of spermatozoa in queens and drones indicate multiple mating in *A. andreniformis* and *A. dorsata*. *Apidologie* 21:281-286.

Koeniger, G., N. Koeniger, H. Pechhacker, F. Ruttner and S. Berg. 1989a. Assortative mating of European honeybees, *A. mellifera ligustica* and *A. mellifera carnica*. *Insectes Soc.* 36:129-138.

Koeniger, G., M. Mardan and F. Ruttner. 1990b. Male reproductive organs of *Apis dorsata*. *Apidologie* 21:161-164.

Koeniger, G., M. Wissel and W. Herth. 1990c. Cornual secretion on the endophallus of the honeybee drone (*Apis mellifera*). *Apidologie* 21:186-191.

Koeniger, N., G. Koeniger, S. Tingek, M. Mardan and T. E. Rinderer. 1988. Reproductive isolation by different time of drone flight between *A. cerana* Fabricius, 1793 and *A. vechti* Maa, 1953. *Apidologie* 19:103-106.

Koeniger, N., G. Koeniger and S. Wongsiri. 1989b. Mating and sperm transfer in *Apis florea*. *Apidologie* 20:413-418.

Koeniger, N. and H. N. P. Wijayagunesekara. 1976. Time of drone flight in the three Asian honeybee species (*Apis cerana, Apis florea, Apis dorsata*). *J. Apic. Res.* 15:67-71.

McEvoy, M. and B. A. Underwood. 1988. The drone and species status of the Himalayan honey bee, *Apis laboriosa* (Hymenoptera: Apidae). *J. Kansas Entomol. Soc.* 61:246-249.

Pain, J. and F. Ruttner. 1963. Les extraits de glandes mandibulaires des reines d'abeilles attirent les males lors du vol nuptial. *C. R. Acad. Sci.* 256:512-515.

Peer, D. F. 1957. Further studies on the mating range of the honeybee *Apis mellifera*. *Can. Entomol.* 89:108-110.

Punchihewa, R. W. K., N. Koeniger and G. Koeniger. 1990a. Mating behaviour of *A. cerana* in Sri Lanka. In *Social insects and the environment, Proceedings of the 11th International Congress of IUSSI.*, G. K. Veeresh, B. Mallik and C. A. Virakthamath, eds. Oxford & IBH Publishing Co. Ltd.: Bombay, Delhi. P. 108.

------. 1990b. Congregation of *Apis cerana indica* Fabricius 1798 drones in the canopy of trees in Sri Lanka. *Apidologie* 21:201-208.

Renner, M. and G. Vierling. 1977. Die Rolle des Taschendrüsenpheromons beim Hochzeitsflug der Bienenkönigin. *Behav. Ecol. Sociobiol.* 2:239-338.

Rowell, G. A., O. R. Taylor Jr. and S. Locke. 1986. Variation in drone mating times among commercial honey bee stocks. *Apidologie* 17:137-158.

Ruttner, H. and F. Ruttner. 1963. Untersuchungen über die Flugaktivität und das Paarungsverhalten der Drohnen. *Bienenvater* 84:297-301.

------. 1966. Untersuchungen über die Flugaktivität und das Paarungsverhalten der Drohnen. 3. Flugweite und Flugrichtung der Drohnen. *Z. Bienenforsch.* 8:332-354.

Ruttner, F. 1988. *Biogeography and Taxonomy of Honeybees.* Springer Verlag: Berlin.

Ruttner, F., J. Woyke and N. Koeniger. 1972. Reproduction in *A. cerana*. 1. Mating behaviour. *J. Apic. Res.* 11:141-146.

Sannasi, A., G. S. Ratulu and G. Sundara. 1971. 9-oxodec-trans-2-enoic acid in the Indian honeybees. *Life Science* 10:195-201.

Shearer, D. A., R. Boch, R. A. Morse and F. M. Laigo. 1970. Occurence of 9-oxodec-trans-2-enoic acid in queens of *A. dorsata, A. cerana* and *A. mellifera. J. Insect Physiol.* 16:1437-1441.

Simpson, H. 1960. Male genitalia of *Apis* species. *Nature* 185:56.

------. 1970. The male genitalia of *Apis dorsata* L. (Hym: Apidae). *Proc. R. Entomol. Soc. Lond. (A)* 45:169-171.

Taber, S. III. 1964. Factors influencing the circadian flight rhythm of drone honey bees. *J. Econ. Entomol.* 48:522-525.

Taylor, O.R. and G. A. Rowell. 1988. Drone abundance, queen flight distance, and the neutral mating model for the honey bee, *Apis mellifera*. In *Africanized honeybees and bee mites.*, Needham, G. R., R. E. Page, M. Delfinado-Baker and C. E. Bowman, eds. Ellis Horwood: Chichester. Pp. 173-183.

Tingek, S., M. Mardan, T. E. Rinderer, N. Koeniger and G. Koeniger. 1988. Rediscovery of *Apis vechti* (Maa, 1953): The Saban honeybee. *Apidologie* 19:97-102.

Tryasko, V. V. 1956. Multiple matings of queen bees. *Pchelovodstvo* 34:29-31 (in russian).

Wongsiri, S., K. Limbipichai, P. Tangkanasing, M. Mardan, T. E. Rinderer, H. A. Sylvester, G. Koeniger and G. Otis. 1990. Evidence of reproductive isolation confirms that *Apis andreniformis* (Smith, 1858) is a separate species from sympatric *Apis florea* (Fabricius, 1787) *Apidologie* 21:47-52.

Woyke, J. 1960. Natural and artificial insemination of queen honeybees. *Pszczelinicze Zeszyty Nankowe* 4:183-275.

------. 1975. Natural and instrumental insemination of *A. cerana indica* in India. *J. Apic. Res.* 14:153-159.

Yoshida, T. and J. Saito. 1990. Mating success of both native *A. cerana japonica* and introduced *A. mellifera* in sympatric condition. In *Social insects and the environment, Proceedings of the 11th International Congress of IUSSI.*, G. K. Veeresh, B. Mallik and C. A. Virakthamath, eds. Oxford & IBH Publishing Co. Ltd.: Bombay, Delhi. Pp. 101-102.

11

Coadaptation of Colony Design and Worker Performance in Honey Bees

Fred C. Dyer

Introduction

Social life is a defining feature of *Apis* biology, and human attempts to understand the honey bee colony, no doubt partly motivated by practical aims, began in antiquity. In the scientific age, the European honey bee, *Apis mellifera*, has come to serve as an important model organism in the study of social behavior in insects (Seeley, 1985, 1989). As a model organism, the honey bee is exemplary. It is interesting, easy to rear, and exceedingly amenable to experimental manipulation. Furthermore, with the exception of its remarkable dance language (Chapter 9, this volume) and its ability to survive winter in a thermoregulating cluster (Seeley, 1985 review), *Apis mellifera* is in many ways a "generic" eusocial insect, exhibiting principles of social organization that recur in many other social species (Wilson, 1971; Seeley, 1985). But as the contributions to this book make abundantly clear, there is more than one way to be a honey bee. Several other *Apis* species live in southern and eastern Asia. All are quite recognizable as honey bees, but they exhibit a number of striking differences -- both socially and individually -- from *Apis mellifera* and from one another. My aim in this chapter is to review two recent investigations of this interspecific diversity which broaden our understanding of honey bee social life, and illuminate some general questions about the ecology and evolution of eusocial behavior in insects.

A major goal of insect sociobiology is to explain the evolution of colonies, which in honey bees, as in other eusocial species, consist of a large number of sterile individuals laboring together in support of the reproduction of the queen, who is ordinarily their mother. This goal encompasses two somewhat narrower (but still broad) sets of questions. First, how and why did social species evolve from solitary ancestors, and what favors the continued subordination of the individual genetic interests of workers to the genetic interests of the queen? Second, in species such as the honey bee, in which the genetic interests of the unreproducing workers are essentially aligned and are almost wholly dependent upon the success of the colony (Seeley, 1989 review), how does natural selection modify worker behavior to result in adaptive colony behavior?

My concern here is with this second question, which can be rephrased by asking how colonies are designed for survival (Wilson, 1985; Seeley, 1989). The answer requires a dual perspective. On one level we need to investigate what traits of the colony as a whole enable it to meet the energetic and nutritional demands of the work force and the developing brood, to survive periods of diminishing food supply, to defend itself against predators, diseases, and climatic fluctuations, and to allocate resources to growth and reproduction. In other words, the colony is treated as a unit of functional design, a "factory enclosed in a fortress" (Oster and Wilson, 1978) whose design properties are shaped by a diverse array of challenges to colony survival. The other level that needs to be investigated is that of the worker. Having identified how the colony responds to a particular challenge, we need to explain the properties of this social response in terms of the properties of its components, the workers, since it is their behavioral, physiological, and morphological characteristics, that natural selection must have shaped to meet the challenge. As Seeley (1989) has pointed out, this dual focus may give insights into one path through which biological complexity evolves, namely through the construction of integrated "survival machines" from components originally designed for independent existence.

The two studies reviewed in this chapter have drawn upon the diversity of social and individual traits among three Asian *Apis* species which exemplify the major contrasts that can be seen in *Apis* biology (Seeley et al., 1982; Dyer and Seeley, 1987, 1991a). Both studies have been carried out in Thailand during extended stays in the same rainforest site, Khao Yai National Park, where all three species live. Both studies have attempted to explain the interspecific differences through joint comparisons of worker traits, social traits, and ecology. Since the three species live in the same range of tropical habitats (Ruttner, 1988), the divergence of traits could be studied in the context of the same array of predators, resources, and climate. The hope of comparative study of this sort (see Bartholomew,

1987) is that the ecological correlates of species differences will suggest hypotheses about the major selective pressures that have led to the diversification, and about the constraints on adaptation imposed by functional interrelationships among traits. The advantage of working on congeneric species is that differences can be attributed to evolutionary modification of a common ancestral groundplan, rather than to different groundplans. What results, then, is an "equilibrium" analysis of functional design (Lauder, 1986; Huey and Bennett, 1987), an attempt to document the different suites of coadapted traits that currently correlate with contrasting challenges to survival faced by colonies of each species. Later in the chapter I will consider how a phylogenetic perspective might affect the insights obtained so far.

The organization of the rest of chapter is as follows. First I briefly review the social biology of *A. mellifera*: what was known prior to the work reviewed here about how the Asian honey bees differed from *A. mellifera*, and some specific questions that this diversity prompted. Then I summarize a study by Seeley et al. (1982) which compared the colony defense strategies of the Asian honey bees. This study has been thoroughly reviewed elsewhere (Seeley, 1983, 1985), so I will keep the present review brief. Then I review a more recent study (Dyer and Seeley, 1987, 1991a) which compared the *Apis* species (including *A. mellifera*) with respect to the energetics of workers and colonies. One theme that unites these studies is that worker design, which is the basis for colony-level adaptive design, is subject to multiple interacting influences, and that these influences can only be appreciated by viewing worker traits in a social context. At the end I outline future directions which I think comparative studies of *Apis* colony design will take.

Ways to Be a Honey Bee Colony
Apis mellifera

The vast literature on *Apis mellifera*'s social organization has been thoroughly reviewed by Michener (1974), Seeley (1985), and Winston (1987). Here I provide a brief synopsis of traits related to defense and energetics. This provides a framework for discussing the diversity among the different *Apis* species. My focus is on the temperate zone European races of *A. mellifera*. For recent reviews of behavioral diversity within *A. mellifera*, see Winston (1987), Ruttner (1988) and Needham et al. (1988).

Apis mellifera colonies consist of some 15,000 workers which nest on a set of several parallel combs suspended from the ceiling of a cavity in a tree or a rock. The cavity walls provide an important line of defense against a wide range of predators and diseases, and against climatic fluctuations. Many larger predators may be totally prevented from reaching the brood, the stored honey and pollen, the wax combs, or the workers themselves; the stinging attacks of the workers punish predators that persist in attacking the cavity walls or the entrance. Defense against disease organisms (viral, bacterial, fungal) is achieved by a variety of measures including keeping the nest clean and removing dead and diseased workers (Seeley, 1985 review).

The cavity walls also provide insulation from cold and shade from the sun, and consequently play an important role in the regulation of the internal climate of the nest. Regulation of temperature and humidity is important for proper larval development, freedom from some diseases, and survival through winter (Seeley, 1985 review). Workers also contribute behaviorally to the regulation of temperature. In hot weather they ventilate the nest by fanning their wings, collect water to spread through the nest, and at the highest ambient temperatures may partially evacuate the nest (Seeley and Heinrich, 1981 review). In cold weather they cluster tightly together in the nest, so restricting the rate at which metabolic heat is lost that the temperature of bees at the core remains >20°C even when ambient temperature falls below 0°C (Heinrich, 1985 review).

As a factory, the colony's task is to fuel its growth, maintenance, and reproduction with nectar and pollen collected from the environment. The foraging strategy of *A. mellifera* is based on a combination of individual and social specializations which result in the efficient discovery and subsequent rapid exploitation of floral resources from a vast area around the nest (Visscher and Seeley, 1982; Seeley, 1985). The individual specializations (some of which may be shared by other bees, including non-social species) include the mechanisms involved in navigation, learning, and making economic decisions (von Frisch, 1967; Dyer and Seeley, 1989). The most important social specialization is the recruiting power of the dance language, which derives from both the spatial precision with which foragers can indicate the location of food they have discovered, and a set of processes that ensure dances are used only for the most profitable patches currently available (Seeley, 1989 review).

The output of the factory are reproductive swarms, each of which consists of 50-60% of the workers from the parent colony. In a northern temperate climate, colonies produce 1 or 2 swarms in late spring, after a period of rapid colony growth (reviews by Seeley, 1985, and Winston, 1987). Apparently the colony needs to delay reproduction long enough to produce a large swarm, yet not delay so long that the daughter colony has insufficient time to store resources needed to survive a difficult winter (Seeley and Visscher, 1985).

In recent years research has turned toward understanding how the activities of workers in an *A. mellifera* colony are coordinated to produce colony traits. Three general organizational principles, which also apply to many other social insect species, have been well documented in honey bees. First, the parallel operation of the workers allows the colony to complete a given task with high efficiency and reliability (Oster and Wilson, 1978); for example the probability that the colony will discover a given patch of food is undoubtedly enhanced by having hundreds of workers searching in parallel (Seeley, 1987). Second, division of labor among the workers means that several different tasks can be performed simultaneously. Labor is divided primarily on the basis of age (Seeley, 1985 review), although an additional genetic component has recently been discovered (Robinson and Page, 1989). Third, the actual coordination of worker behavior is accomplished through the decentralized, rule-based decisions of the individual workers responding to local cues, rather than through either a centralized hierarchy of information flow or global knowledge about colony conditions by any one worker. Particularly clear examples of this principle have been documented for colony thermoregulation (Heinrich, 1985) and colony-level patch choice (Seeley, 1989 review).

Asian Apis: Patterns and Questions

The species that I will discuss are the eastern hive bee, *A. cerana*, the giant bee, *A. dorsata*, and the dwarf bee, *A. florea*. Earlier studies of these species (reviews by Seeley et al., 1982; Ruttner, 1988) described a number of important similarities to *A. mellifera*: all live in large colonies of morphologically identical workers, all build wax combs for rearing brood and storing food, all reproduce by colony fission, all use a dance language, and all share the same habitats over a large region of sympatry in tropical Asia. However, the Asian species also exhibit striking differences, which, with the exception of the dance language (Lindauer, 1956) and certain aspects of the nesting behavior (Morse and Laigo, 1969) were little studied prior to 1975. Three traits in particular seemed to have obvious implications for the ecology of colony design, and consequently were the starting point of the studies discussed here (Table 11.1).

First, worker body size varies over a 5-fold range from the smallest species, *A. florea*, to the largest, *A. dorsata*. Second, colony size (i.e., number of workers) varies over a six-fold range, scaling directly with worker body size. Finally, the Asian species exhibit two distinct types of nest architecture. The nest of *A. cerana* is like that of *A. mellifera* --

Table 11.1 Basic comparison of social and individual differences among four *Apis* species (see Seeley et al., 1982; Seeley, 1985; Dyer and Seeley, 1991a).

	florea	*cerana*	*mellifera**	*dorsata*
Body Mass (mg)	23	44	77	118
Nest Site	Twig in shrub	Cavity	Cavity	Tree limb or cliff
Number of Combs	1	3-8	6-8	1
Colony Population	6,300	9,200	18,800	36,600
Brood:Worker	0.59	1.88	1.77	Not Determined
Cells:Worker	0.80	3.56	3.99	0.74

*Temperate zone race

multiple parallel combs are suspended from the ceiling of a cavity. In both *A. florea* and *A. dorsata*, however, the nest consists of a single comb covered only by a curtain of interlinked workers. Because of the curtain, colonies of both open-nesting species have substantially more adult workers per cell or per brood than do colonies of the cavity-nesting species. Beyond these similarities, the nests of the two open-nesting species differ from each other in a number of respects. *Apis florea*'s small nests are suspended from thin twigs in dense vegation, with wax wrapped around the supporting twig to form a platform on top of the nest; *A. dorsata*'s large combs hang ventrally from overhanging rocks, tree branches, or balconies (see Figure 9.2, this volume).

Each of these traits might be expected to affect both colony defense and colony energetics. Consider defense first. Seeley et al. (1982) realized that a colony's susceptibility to predation would be a function of its risk of being detected, the risk of being approached and attacked once detected, and the risk that the attack would be successful and the nest destroyed. Furthermore, interspecific differences in body size, colony size, and nest architecture (including nesting site) might be correlated with differences in how susceptible colonies are to these risks, and with differences in the array of predators that impose them. For example, Seeley et al. (1982) hypothesized that the exposed nests of *A. dorsata* and *A. florea* should both be more easily approached by some potential predators than the enclosed nests of *A. cerana*, but would differ from each other in how easily their nests could be detected in the first place, and in the potency of the stinging defense. Conceivably, these and other traits might differ among species in part because of adaptive divergence in reponse to predation.

Although Seeley et al. (1982) focused only on predation, they recognized that the energetic performance of workers and colonies would also have to be considered to approach a fuller understanding of the interspecific diversity of worker and colony traits. Dyer and Seeley (1987, 1991a), exploring this aspect of *Apis* ecology, were guided by what seemed to be reasonable assumptions about how the body size of workers would affect their metabolic rate, body temperature, transport costs, load-carrying capacity, flight range, and lifespan. We assumed further that size-dependent scaling of these aspects of forager performance might result in distinct and readily interpretable interspecific trends in how colonies exploit and use resources, including how selective they are, how far their foragers travel, how many patches they exploit simultaneously and how much workers contribute per capita to colony growth. Species with large colony size (and larger workers) should have higher total energy requirements, but if, as we assumed by analogy with other animals (Calder, 1984), they have lower mass-specific metabolic rates, they should use energy more efficiently. Again, these differences might be associated with clear patterns in the collection and allocation of resources. Species differences in nest architecture might result in differences in the energy cost of colony thermoregulation (higher for open-nesters than for cavity nesters, one might surmise), in the nest's per capita food-storing capacity (lower for open-nesters), and in the colony's allocation of workers to foraging instead of nest defense (lower for open-nesters). In summary, it seemed likely that at least some, and possibly many, of the differences among the *Apis* species could be attributed to the evolution of differing strategies for the exploitation and use of resources.

Colony Defense

In an eight-month study in Khao Yai Park, Seeley et al. (1982) compared colony defense strategies of the Asian honey bees with the goal of describing the major natural predators on colonies of each species, and documenting which colony traits may play a role in countering these threats. Their observations suggested that predation has played an important and pervasive role in shaping the social and individual traits of honey bees. Following is a summary of the colony defense strategy of each species as determined by Seeley et al. (1982).

For *A. florea*, the small, exposed nests and the relatively weak stings of the small workers render colonies vulnerable to most predators that discover them. Seeley et al. (1982) observed that *A. florea* nests were more than twice as likely to be destroyed by predators than were nests of the cavity-nester *A. cerana*. The most serious predators of *A. florea* nests were ants, large vespine wasps, tree shrews, and macaque monkeys. A major line of defense against all predators but ants is the crypticity of the nest, which results from the preference of colonies for nest sites in dense vegetation. Of 138 colonies discovered by Seeley et al., 87 (63%) could not be seen from any direction by an observer 5 meters away. Not only do colonies appear to choose concealed nesting sites, but they also tend to desert sites which lose leaf cover during the dry season. By experimentally removing leaves from some nesting sites but not others, Seeley et al. established that colonies were more likely to leave recently exposed sites, even if the nests remained shaded from the sun. One colony that was followed on leaving an exposed site moved to a new site that was totally concealed by dense vegetation. Colonies that departed exposed sites appeared to have prepared for the departure by ceasing brood-rearing several days prior to the move.

If concealment fails and the nest is attacked, the curtain provides a source of defenders. Bees in the curtain of an undisturbed colony are usually quiescent, but they are vigilant and quick to repond to movement near the nest or to mechanical disturbance. An *A. florea* colony is easily overwhelmed, however. The workers may try to mount a defense by either flying off to sting the attacker or, if it is a wasp or an ant, attempting to pull it into the curtain where it can be overpowered. But the attacks of tree shrews and monkeys are probably so sudden and destructive that the colony quickly has no nest left to defend. Wasps (*Vespa tropica*) that Seeley et al. observed attacking *A. florea* colonies appeared to be impervious to the bees' stings, and too strong to overpower. They attacked by moving slowly onto the nest and killing bees as they encountered them; the curtain of bees gradually retreated from the wasps and eventually the colony was driven away, abandoning the brood to the wasps.

Whether an *A. florea* colony departs its nesting site because of loss of cover or an actual attack, it compensates for the energetic costs of renesting by harvesting the wax and honey from the abandoned comb and transporting it to the new site. If the comb is fresh, up to 40% (Dyer, unpublished data) of the wax may be removed and carried like pollen on the workers' hind legs. Since the mass to mass conversion ratio of sugar to wax is roughly 5:1 (Hepburn, 1986), this recycling likely represents a substantial savings of energy that would otherwise have to be collected in the form of nectar.

Though a preference for concealed nest sites may help *A. florea* colonies avoid detection by many predators, the nests are quite detectable by ants, especially the voracious weaver ant (*Oecophylla smaragdina*), which nests and forages in the same sorts of sites where *A. florea* nests. The honey bee's principal defense against these and other ants is to cover the twigs supporting the nest with sticky bands made of plant resins collected by foragers. The colony's preference for attaching the comb to thin twigs probably helps minimize the surface that needs to be defended in this way. The few ants that get past these barriers are usually overwhelmed by bees. Even if denied access to the nest, weaver ant foragers may nevertheless impose a chronic drain on an *A. florea* colony's work force; ants perched on vegetation near the nest can pluck flying bees from the air without assistance.

An *A. cerana* colony is protected from the direct attacks of most predators by the walls of its nest cavity. The nest entrance is small enough to exclude many predators, and is easily defended against predators that might be able to enter. The only predators able to breech these defenses are those powerful enough (such as bears) to enlarge the cavity entrance, or small and numerous enough (such as weaver ants) to overwhelm the workers defending the entrance. In India, for example, I have twice seen large *Oecophylla* colonies drive *A. cerana* colonies from their nests. Still, successful attacks are unusual. The workers behave in ways that enhance the effectiveness of the nest's defenses. When under attack by vertebrates or wasps, they retreat inside rather than, like *A. mellifera*, flying out to sting. Against ants, guard bees can buzz their wings and blow the intruders away with blasts of air. The surfaces around the nest entrance are kept smooth, making it harder for ants to keep their footing. If the attacker is *Oecophylla*, defending bees hover over groups of ants, as if to distract them. Unlike *A. florea*, *A. cerana* workers are too large for *Oecophylla* foragers to handle alone.

The large nests of *A. dorsata* are far more conspicuous than those of the other Asian species. Not only are they established in completely unconcealed sites, but colonies tend to aggregate on certain "traditional" trees, cliffs, or buildings. I once counted 186 colonies on a single banyan tree near Bangalore in southern India. Clearly cypticity plays no part in *A. dorsata*'s defenses against predators. Instead, a colony's defense strategy centers on the inaccessibility of its nesting sites -- tall rock cliffs and the uppermost branches of giant, smooth-barked trees -- and on its violent stinging attacks. As in *A. florea*, the workers forming the curtain are vigilant and easily provoked. Unlike *A. florea*, the defending bees are large enough and numerous enough to punish severely most prospective nest predators. Ants do not pose a problem, perhaps because of the large size of the bees. The threats colonies do face, apart from human beings, for whom wild *A. dorsata* colonies provide a major source of honey in tropical Asia, are specially adapted predators. These include the Eurasian Honey Buzzard and the Malayan honey bear, which both attack *A. dorsata* nests for their brood (Seeley et al., 1982 review). Also, some birds appear to specialize on *A. dorsata* workers coming and going from nests. In India I have observed flocks of Black Drongos (*Dicrurus adsimilis*) and Chestnut-headed Bee-eaters (*Merops leschenaulti*) more or less permanently roosting near aggregations of *A. dorsata* colonies. The birds repeatedly sallied up to intercept flying bees, and occasionally picked bees off the curtain of a colony. Thus they imposed the same sort of chronic losses that weaver ants may impose on an *A. florea* colony.

Table 11.2 summarizes the diversity of colony defense strategies among the Asian bees. Seeley et al.'s (1982) study revealed that interspecific differences in worker and colony traits expose colonies of the three species to different assemblages of predators. Furthermore, the effectiveness of colony defense in each species is concentrated in a different phase of the sequence of events leading to predation. Colonies of *A. florea* are hard for vertebrate and wasp predators to detect but easy for them to approach and consume; ants can detect nests easily and impose a drain on the work force, but not mount a devastating attack. Colonies of *A. cerana* are relatively easy to detect, but fairly secure against attack and consumption. Colonies of *A. dorsata* are easy to detect, but hard to approach and invulnerable to attack by most predators.

One interpretation of these patterns is that predation has played an important selective role in producing at least some of the differences among the *Apis* species. Predation need not have been the initial cause of differences in traits such as body size, colony size, or nest architecture, but it could have favored the exaggeration of differences that arose for other reasons. (Indeed, worker size appears to have been changed more than the size of reproductives; drone mass varies over only a 2-fold range from *A. florea* to *A. dorsata*.) Furthermore, selection pressures associated with predation could have favored the elaboration of traits which enhanced the advantages of a trait such as large size (e.g., aggressiveness in *A. dorsata*) or compensated for the disadvantages of small size (e.g. preference for concealed nesting sites in *A. florea*).

Table 11.2 Comparison of colony defense strategies of three Asian honey bee species.

	A. florea	*A. cerana*	*A. dorsata*
Nest predators	ants, wasps, mammals	ants, mammals, wasps	honey buzzard, humans
Detection	difficult, except for ants	moderately difficult	easy
Approach	easy, except for ants	difficult for most predators	difficult, except for good climbers
Consumption	easy, except for ants	moderately difficult	difficult

Worker and Colony Energetics

With respect to colony defense, the observed ecological correlates of individual and social differences among the Asian honey bees were relatively straightforward to interpret. In the case of the traits related to worker and colony energetics, which we studied during a 10-month stay in Thailand (Dyer and Seeley, 1987, 1991a), a puzzle emerged from the very beginning. We began our study with an investigation of worker body size. Our reasoning was that the differences in body size, whatever the reason they arose, should have pervasive implications for the energetic performance of workers and colonies. This is because mass-specific metabolic rate, which usually scales inversely with body mass in insects and other animals (Bartholomew, 1981; Calder, 1984), should correlate strongly with a variety of other worker traits which we assumed would affect the performance of foragers and consequently their contributions to the energy economy of the colony. We expected to be able to confirm these scaling relationships, and then to investigate how size-dependent constraints on worker performance might correlate with features of colony design involved in the acquisition and allocation of resources. What we found was that traits associated with metabolic rate did not scale with body size in the way we had anticipated. Explaining this puzzling departure became the focus of the study, and led us to surprising insights into the coadaptation of worker performance and colony design.

Worker Energetics

As a starting point we compared the body temperatures of flying bees of each species. In many endothermic insects, including *Apis mellifera* (Heinrich, 1979a), flight muscle temperature is determined largely by a size-dependent balance between the rate of metabolic heat production and the rate of passive heat loss (Bartholomew, 1981). In general, the rate of total heat loss, which is dependent on total surface area, increases with increasing body size, but more slowly than does the rate of total metabolic heat production, which is dependent on flight muscle volume. Consequently larger insects usually produce a larger gradient in flight between thoracic (flight muscle) temperature (T_{th}) and ambient temperature (T_a) than do small insects. This pattern is particularly strong within a group of closely related species (e.g., Bartholmew and Heinrich, 1973, 1978). Hence we assumed that measurements of flight muscle temperature would indicate whether processes of heat loss and heat production scale with body size in honey bees.

To measure body temperature, we used the "grab and stab" technique (Heinrich, 1979b), capturing foragers in flight or shortly after they landed, and quickly inserting a fine thermocouple probe into the thorax. The thorax is the major source of metabolic heat production in flying bees, and the major source of heat loss except at very high ambient temperatures (Heinrich, 1979a,c). We measured T_{th} over a range of T_a. Because the intrinsic rate of heat loss from a body is strongly dependent on T_a, T_{th} should increase with increasing T_a unless actively regulated. In all the Asian *Apis* species, as in other insects of similar body size (May, 1976; Heinrich, 1979a), T_{th} showed this dependence on T_a, so that over a range of T_a from 15-30°C, the gradient T_{th}-T_a remained fairly constant in each species (Figure 11.1). This supported our original assumption that body temperature was the result of a simple balance between the rates of metabolic heat production and convective heat loss.

To our great surprise, however, the Asian honey bees departed from the expected size-dependent trend in body temperature (Figure 11.1). Although the smallest species, *A. florea*, not surprisingly maintained the smallest gradient T_{th}-T_a, the largest species, *A. dorsata*, was consistently cooler in flight than either *A. cerana* or *A. mellifera* (which was previously studied by Heinrich, 1979a). Consistent with these patterns and with previous studies of other insects, the species which maintained larger temperature gradients could keep their flight muscles warm enough to stay in flight at lower T_a than those with smaller gradients (Dyer and Seeley, 1987).

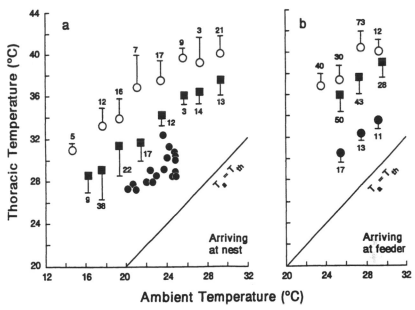

Figure 11.1 Thoracic temperatures (T_{th}) recorded at various ambient temperatures (T_a) in foragers of three Asian *Apis* species caught as as they arrived at (a) their nest or (b) a dish of sugar water to which they had been trained. Filled circles: *A. florea*. Filled squares: *A. dorsata*. Open circles: *A. cerana*. Symbols with error bars give mean ± S.E. (sample size as indicated). Symbols without error bars are separate measurements of *A. florea* workers arriving at their nest.

Confronted with these results, we took a detour to explore some alternative physiological explanations for *A. dorsata*'s unexpectedly low body temperature. One possibility that could be ruled out immediately was that *A. dorsata*'s thoracic mass (hence wing muscle mass) constitutes a disproportionately low fraction of total body mass (Table 11.3). The most plausible remaining hypotheses are (1) that *A. dorsata*'s rate of heat loss at a given ambient temperature is higher than its size relative to *A. cerana* and *A. mellifera* would predict, or (2) that *A. dorsata*'s metabolic rate, and hence its rate of heat production, is disproportionately low. The importance of distinguishing these hypotheses lies in their differing implications for the energy economy of individuals and colonies.

Apis dorsata's rate of heat loss could be disproportionately high either because its intrinsic cooling rate is higher as a result of reduced insulation or disproportionately large surface area, or because for some reason relatively more heat is shunted from the thorax to the head and abdomen than in the smaller species. We excluded the first possibility by recording the cooling rate of freshly killed bees after elevating their body temperatures artificially. Cooling rate scaled inversely with body size in exactly the way one would expect if it were determined primarily by the ratio of surface to volume (see Dyer and Seeley, 1987). We investigated the second possibility by dissecting the circulatory systems of each species. In A. mellifera, warm blood flowing from the thorax into the abodomen transfers much of its heat to cooler blood being pumped anteriorly through the aorta. This countercurrent exchange of heat is enhanced by a series of coils in the segment of the aorta that lies in the petiole, where the warm blood comes into contact with it (Heinrich, 1979a). All three Asian species have very similar coils; A. dorsata's coils show no anatomical evidence that they are designed to allow relatively more heat to escape the thorax, thereby producing low thoracic temperature in flight.

We concluded that A. dorsata's disproportionately low body temperature cannot be attributed to a rate of heat loss that is disproportionately high relative to that of A. cerana and A. mellifera. This leaves the hypothesis that its rate of heat production in flight is disproportionately low, which would suggest that metabolic rate does not scale with body size in the way predicted by theoretical considerations (Calder, 1984) and studies of other insect groups (Bartholomew, 1981). The most direct test of this hypothesis would be a measure of metabolic rate in flying bees, which we did not have the equipment to do. However, three independent lines of indirect evidence suggest that A. dorsata's metabolic rate is in fact disproportionately low. First, its flight speed, which might ordinarily be expected to scale positively with flight muscle mass (Lighthill, 1978), is slower than that of both A. cerana and A. mellifera (Table 11.3), suggesting a disproportionately low power input by the flight muscles. Second, its wing-loading, or the weight supported in flight per unit area of the wings, is lower (Table 11.3). This goes against a well-documented tendency (which has a basis in aerodynamic design principles) for wing-loading to scale with body mass in flying animals (Lighthill, 1978). It is consistent, however, with A. dorsata's lower flight speed and with the notion that power input by its flight muscles is disproportionately low. Aerodynamically and geometrically similar animals with different wing loadings tend to have different mass-specific power requirements for flight (Casey, 1989). For this reason, wing-loading is sometimes a better predictor of body temperature in insects than is body mass (Bartholomew and Heinrich, 1973, 1978; Casey and Joos, 1983). We found this to be the case in the honey bees (Figure 11.2). Third, we calculated for each species the rate of heat production that would be needed to produce the thoracic temperature we measured at a given ambient temperature (Table 11.3), assuming body temperature resulted from simple balance between metabolic heat

Table 11.3 Mass and flight energetics of foraging workers. From Dyer and Seeley (1987, 1991a) and references therein.

	florea	*cerana*	*mellifera*	*dorsata*
Body mass, M_b (mg)*	22.6	43.8	77.2	118.0
Thoracic mass (mg)	7.6	15.3	29.1	45.5
Flight speed (m/s)	4.8	7.3	7.8	7.2
Wing-loading (N/m²)	8.1	13.0	14.2	12.6
Wing-loading x $M_b^{-1/3}$	287.2	369.4	333.7	257.1
Heat-production (W/kg of thorax)	399.6	700.3	642.8	354.7

*Mass of unloaded bee arriving at food.

production and passive heat loss at the cooling rate we had measured (see Dyer and Seeley, 1987). Though such a calculation is subject to many potential errors, we assumed that such errors would be consistent for all the species. The calculation suggested that *A. dorsata*'s mass specific metabolic rate was much lower than would be predicted from its size relative to the other species.

This calculation, as well as a further analysis of wing-loading, also indicated a less obvious, and more surprising result: that *A. dorsata* resembles *A. florea* physiologically in spite of the large difference in body size. Over a wide range of body sizes, mass-specific metabolic rate in animals scales inversely and allometrically with body mass, with a scaling exponent of -0.15 to -0.30 (Bartholomew, 1981). Among the *Apis* we observed no such regular scaling. Instead, the mass-specific rate of metabolic heat production was relatively high in both *A. mellifera* and *A. cerana* and relatively low in both *A. florea* and *A. dorsata* (Table 11.3). In comparisons of wing-loading standardized to body mass we also found *A. dorsata* to resemble *A. florea*, and *A. mellifera* to resemble *A. cerana*. Wing-loading of flying animals and well-designed aircraft generally scales with body mass raised to an exponent of 1/3 (Greenewalt, 1975; Lighthill, 1978). For geometrically similar animals described by the same allometric equation, the ratio of wing-loading to the cube root of body mass should be the same; for Diptera and Hymenoptera the ratio is about 250. Note

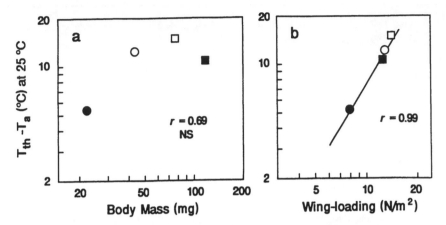

Figure 11.2 The mean temperature gradient measured in bees flying at an ambient temperature of 25°C is plotted as a function of (a) body mass and (b) wing-loading. Open squares: *A. mellifera*; all other symbols as in Figure 11.1. *Apis mellifera*'s temperature data are from Heinrich (1979a), who recorded similar temperature gradients in both African and European races; wing-loading and body mass data are from Dyer and Seeley (1987).

that the value for both *A. florea* and *A. dorsata* is close to 250, but for both *A. cerana* and *A. mellifera* it is considerably above 250 (Table 11.3). One interpretation is that different allometric equations apply to the two pairs of species. Qualitatively, at least, this interpretation also explains interspecific patterns in body temperature, flight speed, and (calculated) mass-specific heat production rate.

Several comparisons of colony-level performance in the Asian species provided further support for the conclusion that the mass-specific expenditure of energy by workers is disproportionately high in *A. cerana* relative to the two other species (Table 11.4). First, measurements of the number of returning foragers per hour at colonies observed on the same day in Khao Yai Park suggested that the foraging activity is relatively intense in *A. cerana*. Far more trips were made in total by the *A. cerana* colony, and the per capita number of trips per day was at least three times as many as for *A. florea* and over 20 times as many as for *A. dorsata*. Second, similar proportions of the work force engage in foraging in *A. cerana* and *A. florea* (*A. dorsata* was not studied in this respect). Therefore, because of the higher number of brood per worker in an *A. cerana* colony (Table 11.1), foragers must provide for the development of three times as many same-sized workers per-capita. Third, colony metabolic rate, as determined in half-day weight-loss experiments, is higher in *A. cerana* than in *A. florea* and *A. dorsata* on a mass-specific and

Table 11.4 Energetics and demography of colonies in Asian honey bees. From Dyer and Seeley (1991a,b).

	florea	*cerana*	*dorsata*
Foraging effort			
Flight range (km)	8	2	10
% of workers foraging	10-23	7-25	ND
Total flights/day	6,400	24,000	2,900
Flights/day/forager	5-10	>20	<1
Colony energy budget*			
Colony biomass (g)	138	423	4,508
Total g sucrose/day	9.3	28.8	78.0
Metabolic rate (mW/bee)	0.29	0.57	0.39
Demography			
Brood:adult worker	0.59	1.88	ND
Maximum lifespan (days)*	>50	<38	ND

*Observations of single colonies
ND - not determined

a per-worker basis. (This contradicted our a priori assumption that mass-specific colony metablic rate would scale inversely with colony biomass.) Fourth, *A. cerana*'s flight range is shorter than *A. florea*'s (Dyer and Seeley, 1991b; see Chapter 9, this volume), which is consistent with it having higher mass-specific transport costs, owing to a higher mass-specific metabolic rate. Finally, *A. cerana*'s lifespan is shorter than *A. florea*'s; this departs from the normal scaling pattern in animals (Calder, 1984 review), but follows the more fundamental trend that longevity scales inversely with mass-specific metabolic rate. (Further supporting the same pattern in honey bees are observations that *A. florea*'s lifespan is about 2.5 times as long as *A. mellifera*'s under common laboratory conditions [Ruttner, 1988].) Any of these results would have been surprising in isolation, but all are consistent with one another and with the original comparisons of body temperature in flight.

I want to stress how striking these patterns are. Allometric scaling relationships are often very successful at describing size-dependent differences among distantly related animal species differing in size by many orders of magnitude. Such relationships are usually even stronger among a group of closely related species. As Calder (1984) has pointed out, departures from a regular scaling trend strongly indicate adaptive modification of the processes that produce such trends. Our results suggest that worker design has diverged in the genus *Apis*, with two species being relatively high-powered in their flight performance and two species being relatively low powered. Any size-dependent trends that may exist are superimposed on this divergence (Dyer and Seeley, 1987). Furthermore, the dichotomy in worker performance parallels the dichotomy in nesting behavior: the high-powered species are cavity-nesters and the low-powered species are open-nesters. The next section explores how this parallel might provide the explanation for the divergence in worker performance.

Colony Design and Worker Tempo

We have interpreted our results in light of Oster and Wilson's (1978) suggestion that the "tempo" of the workers in a social insect colony may evolve in response to ecological pressures operating on the level of the colony. Tempo is defined by Oster and Wilson as a measure of worker performance proportional to the rate at which a worker is likely to accomplish a certain task, and to the energy expended on the task. Worker tempo should affect such aspects of colony performance as how quickly resources are discovered and harvested, how quickly defenders are mobilized to confront a predator, the rate at which workers are accidentally killed, and the degree to which workers waste time or energy by reduplicating one another's efforts. Therefore, as argued by Oster and Wilson, interspecific differences in diet, foraging strategy, resource competition, colony size, worker body size, or susceptibility to predation could lead to the evolution of differences in worker tempo.

It seems appropriate to describe workers of the cavity-nesting species, with their higher mass-specific metabolic rate in flight, higher flight speed, larger number of foraging flights per day, shorter lifespan, heavier work loads in provisioning the brood and building new comb, as being higher in tempo than the workers of the open-nesting species. In any event, tempo can serve as a shorthand term to summarize this suite of apparently correlated worker traits. What then explains the parallel divergence of worker tempo and nest architecture?

Our analysis (for details see Dyer and Seeley, 1991a) focused on the contrast in colony demography between open-nesting and cavity-nesting species, in particular the larger number of workers per cell that the open-nesters apparently need to form a protective curtain. The curtain serves two major functions: protection of the brood comb against predation and thermoregulation at both high and low ambient temperatures. Conceivably, the selective costs or benefits of a particular level of worker tempo could be directly related to the performance of the curtain. The details of how the curtain meets the functions of defense and thermoregulation (Seeley et al., 1982; Heinrich, 1985; Dyer and Seeley, 1987, 1991a; Mardan and Kevan, 1989), however, suggest that it depends not on the tempo of the workers but on their sheer numbers; that is, the curtain of an open-nesting colony would not work as well with a smaller number of bees, whatever their tempo. Hence, we have explored the possibility that worker tempo is set at different levels in open-nesters and cavity-nesters because of its effect on the colony's ability to maintain a given ratio of adults to brood and to exploit the brood-rearing capacity of the nest.

We reasoned that worker tempo should be evaluated in terms of its effects on the material contributions of workers to brood production (including foraging, defense, and work inside the nest) and on worker mortality, which in turn jointly determine the lifetime contributions of workers to colony growth. The essence of our argument is that colonies of the cavity-nesting species would not be able to take full advantage of their capacity for high growth rate if the workers were lower in tempo, and that colonies of the open-nesting species would not be able to produce the demographic structure needed to maintain the curtain if the workers were higher in tempo.

This argument is supported by a consideration of the benefits and costs of worker tempo. We assumed that the rate of energy gained in foraging increases as tempo increases. A higher tempo forager should fly faster between the nest and the food, encounter new flowers more quickly, make more trips per day, and move more quickly in extracting nectar and pollen from a flower. Our data are generally consistent with the hypothesis that *A. cerana* foragers, which we have concluded are higher in tempo than those of the two open-nesting species in the same habitat, collect resources at a faster mass-specific and per capita rate. Additional evidence is provided by observations that *A. cerana* foragers extract food more quickly, and fly more quickly between blossoms, than do either *A. florea* or *A. dorsata* foragers on the same floral species (Murrell and Nash, 1981). Furthermore, *A. cerana*'s higher body temperature allows them to forage at a lower ambient temperature, hence over a wider thermal range. Higher worker tempo might also be beneficial for tasks inside the nest; for example, workers could process incoming food and feed brood at a higher rate, and maintain a larger area of comb per capita.

A major cost of increasing forager tempo would be the associated increase in metabolic costs. Presumably these should rise steadily, whereas improvements in the foraging rate with increasing tempo should rise at a diminishing incremental rate; at some level of tempo, all else being equal, further increases should not improve the foraging rate of workers.

A further cost of increased tempo is its likely effect on worker mortality. In insects and other animals, longevity decreases with increasing mass-specific metabolic rate when different populations are compared, and with increasing activity-dependent metabolic rate within species (Heinrich, 1979a; Neukirch, 1982; Sohal, 1986; Winston, 1987). Among the *Apis* species, longevity, while not correlated with body size as in other animals, is inversely corrleated with mass-specific metabolic rate, which we assume is a fundamental component of worker tempo. Why should increased tempo lead to reduced longevity? For one thing, faster moving insects may be more likely to damage important structures. Also, higher metabolic rates should lead to higher rates of cellular turnover, hence higher rates of fatal DNA copying errors (Kirkwood and Holliday, 1986). Finally, faster moving insects may have less time to detect and avoid predators. Any of these factors could increase the risk of accidental death, and so shorten lifespan.

Consider how these costs and benefits would be balanced differently in open-nesting and cavity-nesting species, because of the permanent presence of the protective curtain in the former but not in the latter. Cavity-nesting species have a much higher potential (per worker) for colony growth because the cavity walls enable a given number of workers to protect and provision a larger number of brood cells. Thus, they have a higher potential to benefit from increased worker tempo. Open-nesting species not only have a lesser capacity for colony growth per capita, hence a lesser potential to benefit if worker tempo is increased, but they might also suffer more from the higher mortality associated with higher tempo. This is because of the more limited capacity of the brood nest to replace adults lost through mortality; higher mortality should translate into fewer adults available to form the curtain and hence poor curtain performance.

We have formalized the assumptions underlying this hypothesis in a simple graphical model (Figure 11.3), which was inspired by a model that Houston et al. (1988) used to study the colony-level consequences of behavioral decisions by foraging honey bees. We started with the assumptions that natural selection maximizes the rate of colony growth, and the per capita growth rate is determined by the per capita rates of worker production and worker mortality, and that production and mortality are both functions of worker tempo. The model compares the effects of increasing tempo on two hypothetical species, one open-nesting and one cavity-nesting. For simplicity, we assumed colony size and worker body size are identical (our data suggest that tempo differences have evolved

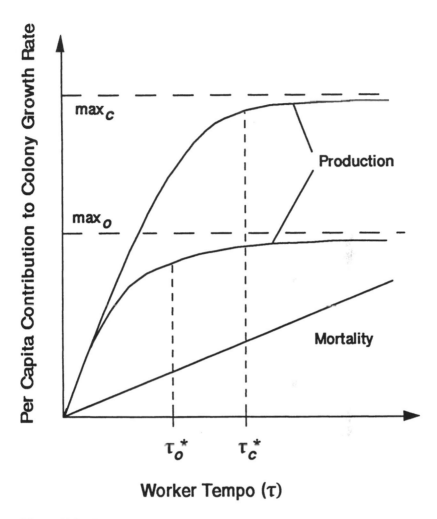

Figure 11.3 Graphical model showing differing optimal tempos for species whose maximum per capita rate of worker production is differently constrained by colony demography associated with nest architecture. The limit on tempo-dependent increases in worker production associated with max_o represents an open-nesting species such as *A. florea* or *A. dorsata*, while max_c represents a cavity-nesting species such as *A. cerana* or *A. mellifera*. Per capita mortality is assumed to increase with (mass-specific) worker tempo in the same way for all species.

somewhat independently of these traits). The per capita contribution to the production of new workers rises with worker tempo in both species. The maximum per capita rate of worker production, however, is higher in the cavity-nester than in the open-nester, because of the larger relative comb area made possible by the protective walls of the cavity. Per capita mortality is assumed to rise with tempo in an identical fashion for the two species, on the grounds that that biological similarities expose them to similar risks of mortality at a given tempo, and endow them with similar capacities to respond to these risks behaviorally and evolutionarily. The model shows that if natural selection favors the level of worker tempo that maximizes the net per capita contribution of workers to colony growth, then the optimal tempo of the open-nesting species should be lower than that of the cavity-nesting species. The model further shows that the per capita mortality rate in the open-nester should evolve to be lower than that in the cavity-nester. Observations of greater lifespan in *A. florea* than in *A. cerana* or *A. mellifera* under similar conditions are consistent with this result.

A further analysis suggests that *A. florea*'s greater lifespan is not only the incidental outcome of selection for colony growth rate, but may itself be of direct selective value because it allows a colony to generate a sufficiently large worker population with a limited comb area. Two general hypotheses could explain the maintenance of a greater number of adult workers per cell in the open-nesting species than in the cavity-nesting species. Either developing workers could pass through the stages from egg to adult more quickly (so adults emerge at a higher rate per cell), or adult lifespan could be longer (so they die at a lower rate per capita relative to the rate of production). Since development times are similar in all four species (18-22 days; see Ruttner, 1988 and Dyer and Seeley, 1991a for sources), longer worker lifespan may be essential to the maintenance of the curtain in open-nesting species. To extend this point, consider how long *A. florea* and *A. cerana* workers must live just to ensure that the colony population stays even, i.e., that old workers are not lost faster than they are replaced through the emergence of pupae. Per capita pupal emergence rate depends upon the development time (approximately 20 days) and on brood population size, which is sharply different relative to adult population size in the two species. Brood population size equals $N \times B$, where N is adult population size and B is the species-specific brood : adult worker ratio (1.88 for *A. cerana* and 0.59 for *A. florea*; Table 11.4). Assuming for simplicity a uniform age distribution in the brood population, then in either species $(N \times B)/20$ of the brood emerges each day. To calculate the average "break-even" lifespan (x, in days), at which adult death rate = pupal emergence rate, we again assume an even age-distribution of adult workers, so $(1/x) \times N$ workers die each day. Setting this death rate equal to the pupal emergence rate, colony size factors out, and $x = 20/B$. Thus for *A. florea*, $x_f = 33.9$ days, and for *A. cerana*, $x_c = 10.6$ days. Note that x_f approximately equals the maximum longevity we observed for *A. cerana* workers (Table 11.4). Thus, *A. florea* colonies, because of the relatively small brood comb available for producing new workers, might actually have trouble growing if worker lifespan were as short as in *A. cerana*.

Viewed from this perspective, *Apis cerana* and *A. mellifera* might be seen as being freed from a demographic constraint on worker lifespan that, in the open-nesting species, is set by the requirement that workers live long enough to staff a populous protective curtain. Any traits (such as those associated with higher worker tempos) which would allow an *A. cerana* (or *A. mellifera*) colony to take advantage of its greater per capita brood-rearing capacity would be heavily favored, even if these traits also tended to shorten worker lifespan.

Whether this analysis is correct in all its particulars, the patterns it attempts to explain underscore an important point which I made at the outset, namely that each trait of individual social insect workers may ultimately be viewed as a social trait, even if its relevance to colony performance is indirect. Our initial assumptions about the relationship between the size and flight performance of workers turned out to be naive precisely on this score. We thought that honey bee workers would be subject to the same physiological constraints as solitary bees of similar size, and that these constraints would in turn constrain the colony's performance in foraging, growth, and reproduction. What we found was that the scaling relationships themselves apparently differ in species with differing nesting behavior.

Ways to Be a Honey Bee Colony: Reprise

The two comparative studies reviewed here together reveal a complex web of influences on the adaptive design of honey bee colonies and workers. The initial study of predation in Thailand by Seeley et al. (1982) had identified three divergent adaptive syndromes, in which worker body size, colony size, and nest architecture seem fundamental in determining the range of predators to which colonies are exposed and the viability of general defensive tactics such as inconspicuousness, inaccessibility, or stinging. Other special traits, such as the selection of nest sites by *A. florea* and *A. dorsata*, the use of sticky bands by *A. florea*, *A. cerana*'s reluctance to leave the nest cavity when under siege, *A. florea*'s tendency to desert a deteriorating nest site, and its recycling of wax after desertion, seem designed either to enhance the effectiveness of general tactics or to minimize susceptibility to particular classes of predators.

The study of worker and colony energetics (Dyer and Seeley, 1987, 1991a) offers a complementary perspective, revealing two divergent adaptive syndromes associated with the divergence in nest architecture (Table 11.5). A colony of a cavity-nesting species has a much higher intrinsic capacity for growth per capita, and the workers are designed to sustain it. The diverse set of worker traits summarized as tempo provide evidence of design for higher work capacity in the cavity-nesting species. Worker performance is at a lower tempo in the open-nesters either because the per capita capacity for colony growth is lower, hence they cannot benefit as much from increased tempo, or because a cost of higher tempo (reduced lifespan) imposes a harsher penalty.

There do appear to be size-dependent trends in worker and colony performance superimposed on this basic divergence between open-nesting and cavity-nesting species. Thus, comparing the two open-nesters, *A. dorsata* foragers have higher flight-muscle temperatures, higher wing-loading, faster flight speed, lower mass-specific metabolic rates, and greater foraging distances than the smaller *A. florea* workers. Similar trends are seen if one compares the two cavity-nesters, *A. mellifera* and *A. cerana*. One reasonable interpretation of these patterns is that open-nesters and cavity-nesters are described by two distinct allometric functions relating worker physiology to body size.

An important link between Seeley et al.'s (1982) study of colony defense and Dyer and Seeley's (1987, 1991a) study of energetics is the emphasis both place on nest architecture, in particular whether colonies nest in cavities or in the open. The colony defense strategies of the two open-nesting species, despite marked differences, both rely heavily on the curtain covering the comb. The curtain also appears to play a an essential role in colony thermoregulation (Dyer and Seeley, 1991a; Mardan and Kevan, 1989). These two functions of the curtain require the maintenance by open-nesting colonies of a large number of adult workers relative to the comb area available to rear them. This demographic shift is responsible for the divergent selection on worker tempo in open-nesting and cavity-nesting species.

Future Directions

Despite the advances made, large gaps remain in our understanding of the diversity of worker and colony design in *Apis*. Following is my list of some major questions that need to be addressed.

Table 11.5 Summary of individual and social patterns in energetics of *Apis*

	Open-Nesting (Low Tempo)	Cavity-Nesting (High Tempo)
Brood:Worker Ratio	Low	High
Mass-Specific Metabolic Rate	Low	High
Per Capita Work Load	Low	High
Potential Colony Growth Rate (Per Capita)	Low	High
Lifespan	Long	Short

Energetics and Foraging Ecology

The studies by Dyer and Seeley (1987, 1991a) have barely scratched the surface of the problem of how body size, colony size, and nest architecture interact to constrain the energetic performance of individuals and colonies. Although Dyer and Seeley (1987) have established that size by itself does not explain the interspecific differences in worker physiological performance, the differences that are observed likely have important implications for foraging energetics. For example, species in which the foragers have lower total metabolic rates can break even energetically on nectar of lower quality, hence, all else being equal, might be less selective about the resources they exploit. Of course, other traits besides worker metabolic rate differ among species, making quantitative predictions difficult; most important, since honey bee foragers collect resources not just to meet their own energy needs but are foraging for the colony, colony size (which sets the total colony metabolic rate) and nest architecture (which influences the amount of excess resources stored per forager) should both determine whether a given resource is economical to harvest. The point, however, is that it is reasonable to expect significant interspecific differences in the economics of foraging, and possibly in the exploitation strategies of workers and colonies. Hence further comparative studies might complement and extend the growing body of work on how *Apis mellifera* foragers and colonies are designed for economic exploitation of energy resources (e.g., Schmid-Hempel et al., 1985; Seeley, 1986, 1989; Wolf and Schmid-Hempel, 1990).

Data that allow these sorts of analyses would also allow us to address various other issues. For example, a particularly intriguing question is how the Asian honey bee species interact with one another in the same habitat. Since the bees do share the same range of habitats, and since their dances probably endow colonies of all three species with tremendous power to exploit floral resources, there is at least the potential for competitive interactions. Indeed, Koeniger and Vorwohl (1979) documented considerable overlap among the three *Apis* species in Sri Lanka in the pollen species present in their nests. Although Koeniger and Vorwohl observed aggressive interactions among species at a dish of sugar syrup, there is no evidence that aggression is an important component of the colonial foraging strategy of any *Apis* species, in contrast to some species of stingless bees (e.g., Johnson and Hubbell, 1974, 1975). Instead, any competitive interactions are likely to be indirect, mediated by the impact of colonies upon available floral resources. Furthermore, understanding these interactions will require taking account of species differences in the demand of colonies for resources, in the thoroughness with which resources are exploited, and in the efficiency with which they are used -- traits at least partly influenced by body size, colony size, and nest architecture. A model for this type of investigation would be the comparison by Schaffer et al. (1979) of competition for a common floral resource by three bee species differing in body size and social complexity.

Finally, I will mention three other problems that might be understood through further investigations of worker and colony foraging ecology:

1. the ecological consequences of *A. dorsata*'s tendency to aggregate large numbers of colonies in a single location.
2. the energetics of the seasonal long-distance migrations which are undertaken by *A. dorsata* but not by *A. cerana* or *A. florea* (Koeniger and Koeniger, 1980).
3. the roles played by different *Apis* species in the pollination of forest plants.

Reproductive Strategies

Apart from some limited information on the mating biology of the Asian *Apis* species (reviewed by Ruttner, 1988 and Chapter 10, this volume), very little is known about the reproductive strategies of colonies of these species. There is ample reason to expect some diversity. Swarming frequency should be affected by per capita colony growth rate, hence by the foraging performance of workers and by the brood-rearing capacity of the nest, traits which show marked interspecific differences associated with nest architecture. But the size of the swarms produced would also affect how frequently they could be produced, and adaptive swarm size might differ among species depending upon the factors determining swarm survival.

As reviewed by Ruttner (1988), there is evidence of interspecific differences in *Apis* swarming biology, but conflicting observations about what the typical swarming pattern of each species is. Apparently, *A. cerana* resembles *A. mellifera* in producing one or two large swarms after a period of rapid colony growth. Colonies of *A. florea*, by contrast, have been reported to produce a series of 5 or more small swarms, culminating in the dissolution of the parent colony (Lindauer, 1956), but other observers have reported that the parent colony may survive and continue to grow (see Ruttner, 1988). In India, I observed (unpublished) an *A. florea* colony produce at least 4 very small swarms in succession, then thrive for several more months, until it was blown down by a storm. Swarming in *A. dorsata*, as in *A. florea*, has been reported to involve the production of a series of small swarms spaced at 3-4 day intervals (Lindauer, 1956), but other patterns are also observed. During a study in India, one *A. dorsata* colony which I was studying produced a single large swarm which depleted the worker population at least by half. Lindauer (1956) described yet another mode of "swarming" in which workers from a large *A. dorsata* colony walked a short distance from the parent nest, began building a new comb, and eventually founded an independent colony on it. One interpretation of these various observations is that patterns of swarm production not only differ among species, but also vary within species (in *A. dorsata* and *A. florea*), either regionally or seasonally.

Other Species and Subspecies of Apis

This review has focused on the diversity in colony design among the three Asian *Apis* species that have been most thoroughly studied. Investigations of other populations have uncovered evidence of still more diversity. Some is seen in the form of specialized traits associated with extreme habitats. For example, *A. laboriosa*, a high altitude (Himalayan) bee with nesting behavior similar to *A. dorsata*, exhibits a distinctive set of traits apparently designed for winter survival (Underwood, 1990). Colonies migrate to lower altitudes and essentially hibernate, building no comb, slowing down their metabolism, and foraging only to meet basic energetic needs. When spring comes they migrate back to high-altitude nesting sites and foraging habitats. Another example is an interesting population of *A. florea* that lives around the Persian Gulf (reviews by Free, 1982; Ruttner, 1988). In studies in Oman, these bees were observed to live in cavities, though they build only a single comb and cover it with a curtain of workers, as colonies in other *A. florea* populations do. Workers and colonies of *A. florea* in this region are also considerably larger than those in tropical Asia. Further studies of these ecologically specialized populations might help illuminate the malleability of worker and colony traits.

Otis (Chapter 2, this volume) reviews what is known about the biology of other *Apis* populations in Asia, including morphotypes which have recently been recognized as full species. Biologically and ecologically, each of these other species closely resembles one of the species I have discussed. Clearly, however, further study of worker and colony traits is warranted, especially where two similar species occur in sympatry and might be expected to have diverged. Examples include the hive bees *A. cerana* and *A. koschevnikovi*, and the dwarf bees *A. florea* and *A. andreniformis* (see Chapter 2, this volume, for details).

Phylogenetic Analysis

The two studies in Thailand that have been the main concern of this chapter consisted of analyses of the current adaptive significance of traits that differ among the living *Apis* species, and made no attempt to consider the historical events leading to these differences. A historical perspective based on the phylogeny of a group under study can be useful for analyses of adaptive function if it establishes the ancestral state of a given homologous trait which varies among species. Polarizing alternative character states can in turn suggest more specific hypotheses about the selective pressures that led to the modification of the trait than if the polarity of change remains ambiguous (Coddington, 1988).

The starting point for a phylogenetic analysis of honey bee colony design has to be a hypothesis about *Apis* phylogeny based on independent characters; none was available until recently, but now Alexander (1991 and Chapter 1, this volume) has proposed one based on a study of morphology. To polarize behavioral character states within the *Apis*, one may also need to examine homologous characters in an outgroup comparison. A problem here is that the choice of the outgroup for honey bees is unclear because the phylogeny of the family Apidae (which includes stingless bees, bumble bees, and orchid bees) is still unresolved (Chapters 3, 4 and 5, this volume).

Nevertheless, certain questions can be addressed using Alexander's cladogram, which is consistent with a phylogeny originally proposed by Lindauer (1956) based on behavioral characters (see Chapter 9, this volume). In cladistic notation, the species I have been discussing are related as follows: *florea* + (*dorsata* + (*cerana* + *mellifera*). This means that, morphologically, *A. florea* departs least among the honey bees from the bee taxa used as the outgroup, in that it has the greatest tendency to exhibit primitive states of the characters studied. The other honey bees form a cluster in which *A. dorsata* departs least from *A. florea*. (Alexander also studied *A. andreniformis*, a dwarf bee that turned out to be a sister species to *A. florea*, and *A. koschevnikovi*, an Asian hive bee that is a sister species to *A. cerana*.) Turning to social behavior, this cladogram suggests that the common ancestor of the extant honey bee species nested on an exposed comb, and that cavity-nesting is a derived condition that arose in the common ancestor of *A. cerana* and *A. mellifera*. As noted by Koeniger (1976) and Alexander (1991), no other bees in the Apidae live on exposed nests, and so the nest architecture of *A. florea* and *A. dorsata* is derived on some level; assuming that the ancestral *Apis* was cavity-nesting, however, would imply that the complex suite of traits shared by the open-nesting species evolved in parallel in dwarf bees and giant bees. Furthermore, apart from being constructed in cavities, the nest architecture of the cavity-nesting *Apis* species shows little resemblance to that of other social bees (Michener, 1974), hence there is little to support the hypothesis that cavity-nesting is the ground-plan state for honey bees.

If we can assume that open-nesting is primitive, then we can draw some conclusions about the polarity of other traits that are correlated with the fundamental contrast in nest architecture between open-nesting and cavity-nesting species. In particular, worker tempo must have been relatively low in the primitive condition, and then increased in association with the move into cavities by the common ancestor of *A. cerana* and *A. mellifera*. For example, perhaps a move into cavities was initially favored by climatic or predation pressures, and was followed by changes in the pattern of comb construction, a shift in colony demography (i.e., increased brood to worker ratio), and ultimately by an increase in worker tempo as the pressure to devote labor to the protective curtain was reduced and the capacity for per capita colony growth was increased. An alternative possibility is that worker tempo changed first and was followed by a move into cavities and a change in nest architecture; this seems less likely, however, since high worker tempo would have to arise in a context (a species with open-nesting colonies) in which, as argued above, it should be disfavored.

To extend this sort of phylogenetic analysis to other aspects of worker and colony design will require considerably more information than we currently have about the social biology of the honey bee species other than *Apis mellifera*, and of other social bees.

Relevance to Other Social Insects

The research in Thailand by Seeley et al. (1982) and Dyer and Seeley (1987, 1991a, b) has not only provided new insights and raised new questions about the diversity of honey bee biology, but has also offered the following general perspectives on the study of individual and colony design in social insects. First, these studies taken together underscore the point (previously emphasized by Oster and Wilson, 1978, and Wilson, 1985) that hypotheses about the adaptive design of colonies have to consider multiple aspects of colony performance, including (at least) defense, energetics, and reproduction. Though a full understanding of *Apis* is still to be reached, this integrated approach has already shown that the functional role of some traits can be appreciated only in such a broad framework, and has identified suites of apparently coadapted traits (e.g., nest architecture, allocation of labor to defense, colony demography, and various aspects of worker flight performance) whose functional interrelationships had not been immediately obvious.

Second, the study by Dyer and Seeley (1987, 1991a) provides considerable support for Oster and Wilson's (1978) suggestion that adaptive worker tempo constitutes an important aspect of colony design. Since this concept was first advanced, only a few studies have explored it empirically (e.g., Franks, 1985; Leonard and Herbers, 1986). Even if the factors that shape worker tempo in other social insects turn out to be different from those proposed for honey bees, the patterns in *Apis* show that tempo may help explain the ecological significance of a diverse set of worker and colony traits.

Finally, the studies in Thailand underscore the value of detailed ecological and behavioral comparisons of a group of closely related species (cf. Bartholomew, 1987). The patterns described in this chapter, and the lessons just mentioned, arguably might never have emerged in studies of a single species, or even in cursory comparisons of the gross differences among the species. For example, the association between nest architecture and worker tempowas detected only after our assumptions about the scaling of worker traits with body size were tested and found to be incorrect. The potential of such comparisons is far from exhausted for honey bees; indeed, one major source of insights, the historical perspective that a phylogenetic analysis would provide, is virtually untapped. Other groups of social insects offer similar unexploited potential. The stingless bees (Meliponinae) in particular, which exhibit an even greater diversity than *Apis* in colony biology, and yet are much less thoroughly investigated (Roubik, 1989 review) represent a particularly attractive frontier.

Acknowledgments

The author's research was funded by NSF grant BNS8405962 to Yale University, and was carried out in collaboration with T. D. Seeley.

Literature Cited

Alexander, B. A. 1991. Phylogenetic analysis of the genus *Apis* (Hymenoptera: Apidae). *Ann. Entomol. Soc. Am.* 84:137-149.

Bartholomew, G. A. 1981. A matter of size: an examination of endothermy in insects and terrestrial vertebrates. In *Insect thermoregulation.*, B. Heinrich, ed. Wiley & Sons: New York. Pp. 45-78.

------. 1987. Interspecific comparison as a tool for ecological physiologists. In *New directions in ecological physiology.*, M. E. Feder, A. F. Bennett, W. W. Burggren, and R. B. Huey, eds. Cambridge University Press: Cambridge. Pp. 11-35.

Bartholomew, G. A. and B. Heinrich. 1973. A field study of flight temperatures in moths in relation to body weight and wing loading. *J. Exp. Biol.* 58:123-135.

------. 1978. Endothermy in African dung beetles during flight, ball making, and ball rolling. *J. Exp. Biol.* 73:65-83.

Calder, W. A. 1984. Size, function, and life history. Harvard University Press: Cambridge, MA.

Casey, T. M. 1989. Oxygen consumption during flight. In: *Insect flight.*, G. J. Goldsworthy and C. H. Wheeler, eds. CRC Press: Boca Raton, Florida. Pp. 257-272.

Casey, T. M. and B. A. Joos. 1983. Morphometrics, conductance, thoracic temperature, and flight energetics of Noctuid and Geometrid moths. *Physiol. Zool.* 56:160-173.

Coddington, J. A. 1988. Cladistic tests of adaptational hypotheses. *Cladistics* 4: 3-22.

Dyer, F. C. and T. D. Seeley. 1987. Interspecific comparisons of endothermy in honey bees (*Apis*): deviations from the expected size-related patterns. *J. Exp. Biol.* 127:1-26.

------. 1989. Orientation and foraging in honey bees. In *Insect flight.*, G. J. Goldsworthy and C. H. Wheeler, eds. CRC Press: Boca Raton, FL. Pp. 205-230.

------. 1991a. Nesting behavior and the evolution of worker tempo in four honey bee species. *Ecology* 72:156-170.

------. 1991b. Dance dialects and foraging ranges in three Asian honey bee species. *Behav. Ecol. Sociobiol.* (in press).

Franks, N. R. 1985. Reproduction, foraging efficiency and worker polymorphism in army ants. In *Experimental behavioral ecology and sociobiology.*, B. Hölldobler and M. LIndauer, eds. G. Fisher Verlag: Stuttgart (*Forschritte der Zoologie* 31). Pp. 91-107.

Free, J. B. 1982. The biology and behaviour of the honeybee *Apis florea*. In *Social insects in the tropics.*, P. Jaisson, ed. Université Paris-Nord. Pp. 181-187.

Frisch, K. von. 1967. *The dance language and orientation of bees.* Belknap/Harvard University Press: Cambridge, MA.

Greenewalt, C. H. 1975. The flight of birds. The significant dimensions, their departures from the requirements for dimensional similarity, and the effect on flight aerodynamics of that departure. *Trans. Am. Phil. Soc.* 65:1-67.

Hepburn, H. R. 1986. *Honeybees and Wax.* Springer Verlag: Berlin.

Heinrich, B. 1979a. Thermoregulation of African and European honeybees during foraging, attack, and hive exits and returns. *J. Exp. Biol.* 80:217-229.

------. 1979b. *Bumblebee economics*. Belknap/Harvard University Press: Cambridge, MA.

------. 1979c. Keeping a cool head: honeybee thermoregulation. *Science* 205:1269-1271.

------. 1985. The social physiology of temperature regulation in honeybees. In *Experimental behavioral ecology and sociobiology.*, B. Hölldobler and M. Lindauer, eds. G. Fisher Verlag: Stuttgart *(Fortschritte der Zoologie* 31) Pp. 393-406.

Houston, A., P. Schmid-Hempel, and A. Kacelnik. 1988. Foraging strategy, worker mortality, and the growth of the colony in social insects. *Am. Nat.* 131:107-114.

Huey, R. B. and A. F. Bennett. 1987. Phylogenetic studies of coadaptation: preferred temperatures versus optimal performance temperatures of lizards. *Evolution* 41:1098-1115.

Johnson, L. K. and S. P. Hubbell. 1974. Aggression and competition among stingless bees: field studies. *Ecology* 55:120-127.

------. 1975. Contrasting foraging strategies and coexistence of two bee species on a single resource. *Ecology* 56:1398-1406.

Kirkwood, T. B. L. and R. Holliday, 1986. Ageing as a consequence of natural selection. In *The biology of human ageing.*, A. H. Bittles, and K. J. Collins, eds. Cambridge University Press: Cambridge. Pp. 1-16.

Koeniger, N. 1976. Neue Aspekte der Phylogenie innerhalb der Gattung *Apis. Apidologie* 7:357-366.

Koeniger, N. and G. Vorwohl. 1979. Competition for food among four sympatric species of Apini in Sri Lanka *(Apis dorsata, Apis cerana, Apis florea,* and *Trigona iridipennis). J. Apic. Res.* 18:95-109.

Koeniger, N. and G. Koeniger. 1980. Observations and experiments on migration and dance communication of *Apis dorsata* in Sri Lanka. *J. Apic. Res.* 19:21-34.

Lauder, G. V. 1986. Homology, analogy, and the evolution of behavior. In *Evolution of animal behavior: paleontological and field approaches.*, M. H. Nitecki and J. A. Kitchell, eds. Oxford University Press: Oxford. Pp. 9-40.

Leonard, J. G. and J. M. Herbers. 1986. Foraging tempo in two woodland ant species. *Anim. Behav.* 34:1172-1181.

Lighthill, J. 1978. Introduction to the scaling of aerial locomotion. In *Scale effects in animal locomotion.*, T. J. Pedley, ed. Academic Press: London, New York. Pp. 365-404.

Lindauer, M. 1956. Über die Verstendigung bei indischen Bienen. *Z. vgl. Physiol.* 38:521-557.

Mardan, M. and P. G. Kevan. 1989. Honeybees and "yellow rain". *Nature* 341:191.

May, M. L. 1976. Thermoregulation and adaptation to temperature in dragonflies (Odonata: Anisoptera). *Ecol. Monogr.* 46:1-32.

Michener, C. D. 1974. *The social behavior of the bees.* Harvard University Press: Cambridge, MA.

Morse, R. A. and F. M. Laigo. 1969. *Apis dorsata* in the Philippines. *Monographs of the Philippine Association of Entomology* 1:1-96.

Murrell, D. C. and W. T. Nash. 1981. Nectar secretion by *toria (Brassica campestris* L. v. *toria)* and foraging behaviour of three *Apis* species on toria in Bangladesh. *J. Apic. Res.* 20:34-38.

Needham, G. R., R. E. Page, Jr., M. Delfinado-Baker, and C. E. Bowman (eds.). 1988. Africanized honey bees and bee mites. Ellis Horwood: Chichester.

Neukirch, A. 1982. Dependence of the life span of the honeybee *(Apis mellifica)* upon flight performance and energy consumption. *J. Comp. Physiol.* 146:35-40.

Oster, G. F. and E. O. Wilson. 1978. *Caste and ecology in the social insects.* Princeton University Press: Princeton, NJ.

Robinson, G. and R. E. Page, Jr. 1989. Genetic determination of nectar foraging, pollen foraging, and nest-site scouting in honey bee. *Behav. Ecol. Sociobiol.* 24:317-323.

Roubik, D. W. 1989. *Ecology and natural history of tropical bees.* Cambridge University Press: Cambridge.

Ruttner, F. 1988 *Biogeography and taxonomy of honey bees.* Berlin-Heidelberg-New York: Springer-Verlag.

Schaffer, W. M., D. B. Jensen, D. E. Hobbs, J. Gurevitch, J. R. Todd, and M. V. Schaffer. 1979. Competition, foraging energetics, and the cost of sociality in three species of bees. *Ecology* 60:976-987.

Schmid-Hempel, P., A. Kacelnik and A. I. Houston. 1985. Honeybees maximize efficiency by not filling their crop. *Behav. Ecol. Sociobiol.* 17:61-66.

Seeley, T. D. 1983. The ecology of temperate and tropical honeybee societies. *Am. Sci.* 71: 264-272.

------. 1985. *Honeybee ecology. A study of adaptation in social life.* Princeton University Press: Princeton, NJ.

------. 1986. Social foraging by honeybees: how colonies allocate foragers among patches of flowers. *Behav. Ecol. Sociobiol.* 19: 343-354.

------. 1987. The effectiveness of information collection about food sources by honey bee colonies. *Anim. Behav.* 35:1572-1575.

------. 1989. The honey bee colony as superorganism. *Am. Sci.* 77:546-553.

Seeley, T. D. and B. Heinrich. 1981. Regulation of temperature in the nests of social insects. In *Insect thermoregulation.*, B. Heinrich, ed., Wiley and Sons: New York. Pp. 159-234.

Seeley, T. D., R. H. Seeley, and P. Akratanakul. 1982. Colony defense strategies of the honeybees in Thailand. *Ecol. Monogr.* 52:43-63.

Seeley, T. D. and P. K. Visscher. 1985. Survival of honeybees in cold climates: the critical timing of colony growth and reproduction. *Ecol. Entomol.* 10:81-88.

Sohal, R. S. 1986. The rate of living theory: a contemporary interpretation. In *Insect aging.*, K.-G. Collatz and R. S. Sohal, eds. Springer Verlag: Berlin. Pp. 23-44.

Underwood, B. 1990. Seasonal nesting cycle and migration patterns of the Himalayan honey bee *Apis laboriosa*. National Geographic Research 6:276-290.

Visscher, P. K. and T. D. Seeley. 1982. Foraging strategy of honeybee colonies in a temperate deciduous forest. *Ecology* 63:1790-1801.

Wilson, E. O. 1971. *The insect societies.* Belknap/Harvard University Press: Cambridge, MA.

------. 1985. The principles of caste evolution. In *Experimental behavioral ecology and sociobiology.*, B. Hölldobler and M. Lindauer, eds. G. Fisher Verlag: Stuttgart (*Fortschritte der Zoologie* 31). Pp. 307-324.

Winston, M. L. 1987. *The biology of the honey bee.* Harvard University Press: Cambridge.

Wolf, T. J. and P. Schmid-Hempel. 1990. On the integration of individual foraging strategies with colony ergonomics in social insects: nectar-collection in honeybees. *Behav. Ecol. Sociobiol.* 287:103-111.

About the Book and Editor

The honey bees were formerly seen as a small, static group comprising four species, whose behavior and ecology were simple variants on the patterns found in *Apis mellifera*. The picture now is one of a large, actively speciating group, reflecting in part the complex geological and biological influence of the *Apis* environment. Research on this diversity has benefitted from new techniques of DNA analysis applied to several long-standing problems in honey bee phylogenetics and that are reported in this volume. The behavior and ecology of the *Apis* species and populations are also more diverse and differentiated than previously recognized: Radically different orientation systems as expressed through dance language exist in various species. This study of *Apis* will be of great interest not only to biologists and apiculturalists but to anyone interested in systematics, genetics, and ethology.

Deborah Roan Smith is assistant professor in the Department of Entomology at the University of Kansas. She was formerly research scientist at the Laboratory for Molecular Systematics, Museum of Zoology, University of Michigan.

247

Author Index

Page citations for senior authors are given in bold-faced type.

Subject Index

Printed and bound by CPI Group (UK) Ltd, Croydon, CR0 4YY

23/10/2024

01778240-0012